SpringerBriefs in Molecular Science

Electrical and Magnetic Properties of Atoms, Molecules, and Clusters

Series editor

George Maroulis, Patras, Greece

More information about this series at http://www.springer.com/series/11647

Victor N. Cherepanov · Yulia N. Kalugina
Mikhail A. Buldakov

Interaction-induced Electric Properties of van der Waals Complexes

 Springer

Victor N. Cherepanov
Department of Physics
Tomsk State University
Tomsk
Russia

Yulia N. Kalugina
Department of Physics
Tomsk State University
Tomsk
Russia

Mikhail A. Buldakov
Institute of Monitoring of Climatic and
  Ecological Systems of the Siberian
  Branch of the Russian Academy of
  Science
Tomsk
Russia

ISSN 2191-5407   ISSN 2191-5415   (electronic)
SpringerBriefs in Molecular Science
ISSN 2191-5407   ISSN 2191-5415   (electronic)
Electrical and Magnetic Properties of Atoms, Molecules, and Clusters
ISBN 978-3-319-49030-4   ISBN 978-3-319-49032-8   (eBook)
DOI 10.1007/978-3-319-49032-8

Library of Congress Control Number: 2016956627

Printed on acid-free paper

This Springer imprint is published by Springer Nature
The registered company is Springer International Publishing AG
The registered company address is: Gewerbestrasse 11, 6330 Cham, Switzerland

Preface

The electrical properties of molecules and molecular complexes are the subject of a close attention up today. While the multipole electric moments have studied good enough, any information about the polarizabilities, especially higher polarizabilities, is quite scarce. At the same time, these interaction-induced properties play often the key role to understand many physical and chemical properties of interacting molecules or atoms. The knowledge of such interactions is important for lot of applications in science and engineering. This small book summarizes mainly the researches carried out last decade by group of Tomsk State University and devoted to interaction-induced electrical properties (multipole moments, (higher)polarizabilities) of van der Waals complexes. We hope that this short survey gives useful contribution to the theory of molecular interactions to be applied for modeling of different physical chemical properties of interacting species, for describing the long-range potential energy surfaces and the surfaces of some electrical properties of van der Waals complexes.

We want to express our appreciation to one of the authors of the book, to Dr. Mikhail Buldakov, who being the talented experimentalist was also often as a mover for many of theoretical researches discussed in the book. Unfortunately, he passed away when the book was still far to be finished. But the main ideas of the book were agreed with him. Moreover, we did not consider it possible to do large changes in the texts of original papers which were used to be included to the book because the results underlying these articles have been suffered in the heated arguments with him and bear his spirit.

Also, we thank Dr. Natalia Zvereva-Loëte and Dr. Vincent Boudon for joint efficient researches of the interaction energies, dipole moments, and polarizabilities of the van der Waals complexes CH_4–N_2 and C_2H_4–C_2H_4, the results of which were also used in the book.

Tomsk, Russia

Victor N. Cherepanov
Yulia N. Kalugina

Contents

Chapter 1
Introduction

Van der Waals complexes. Well-known, that all atoms and molecules can interact together to form either a new molecule or a molecular cluster [1–10]. Sometimes say that such molecular clusters are defined by noncovalent interactions or by van der Waals (vdW) interactions. In this book the term Van der Waals complexes (or clusters) will be used both to a transient collisional complex and to a weakly bound van der Waals molecule. Note that their electric and mechanical properties are not the simple sum of similar properties of components forming them. Other characteristic feature of the van der Waals complexes is their nonrigid structure [1, 2]. As dissociation energy of the van der Waals complexes is much lower of the dissociation energy for usual molecules, such complexes are easily broken up. For this reason, the quantity of the van der Waals complexes in a gas under normal conditions is rather insignificant. However, being present in all gases, the complexes appreciably influence on their properties and take part in all physical and chemical processes of gas media (atmospheres of Earth and other planets, interstellar medium and so on). Furthermore, weakly bound molecular complexes are of great importance for studying the nature of nonbonding interactions that play an important role in many fields of chemistry and solid state physics. Nonbonding interactions of *p*-systems control structures and properties of organic molecules in condensed phase form molecular associates in fluids and play a key role in biomolecules such as nucleic acids, proteins, DNA and RNA [4–6]. As a result, the van der Waals complexes are intensively studied in the last decades. And in the book some aspects of electrical properties of van der Waals complexes, which are not delighted enough in scientific literature, are discussed.

Classification of intermolecular interactions. In this small book the two-body interactions, which are the strongest, are only considered for study the electrical properties of van der Waals complexes. Particularities of many-body interactions may be found, for example, in [11, 12]. Usually, the intermolecular interactions, the physical nature of which is electromagnetic one, are classified according to the distance R between interacting molecules (or atoms). They are the long-range interactions when the interaction energy has a form $U \sim 1/R^n$ and the short-range

© The Author(s) 2017
V.N. Cherepanov et al., *Interaction-induced Electric Properties of van der Waals Complexes*,
SpringerBriefs in Electrical and Magnetic Properties of Atoms, Molecules, and Clusters,
DOI 10.1007/978-3-319-49032-8_1

interactions $(U \sim \exp(-\alpha R))$. The main long-range contributions to the energy and electric properties of interacting molecules are as follows: electrostatic (multipole-multipole), polarization (induction and dispersion), resonance, relativistic, magnetic and retardation. The short-range interactions include the direct electrostatic, exchange, repulsion, change transfer effects. Note that the long-range interactions can be well described in an analytical form using classical electrodynamics for some approximation of the perturbation theory. In turn, for short distances the methods of quantum chemistry are only effective. In part of analytical theory for large R this book will only focus on interaction-induced theory.

References

1. A. van der Avoid, P.E.S. Wormer, R. Moszynski, From intermolecular potentials to the spectra of van der Waals molecules, and vice versa. Chem. Rev. **94**, 1931–1974 (1994)
2. P.E.S. Wormer, A. van der Avoid, Internuclear potentials, internal motions, and spectra of van der Waals and hydrogen-bonded complexes. Chem. Rev. **100**, 4109–4143 (2000)
3. G. Chalasinski, M.M. Szczesniak, State of the art and challenges of the ab initio theory of intermolecular interactions. Chem. Rev. **100**(1), 4227–4252 (2000)
4. K. Müller-Dethlefs, P. Hobza, Noncovalent interactions: a challenge for experiment and theory. Chem. Rev. **100**(11), 143–167 (2000)
5. B. Pullman (ed.), *Molecular associations in biology* (Academic Press, New York, 1968)
6. A.D. Buckingham, in *Intermolecular Interaction: From Diatomic to Biopolymers*, ed. by B. Pullman (Wiley, New York, 1978), pp. 1–68
7. S. Kielich, *Molekularna Optyka Nieliniowa (Nonlinear Molecular Optics)* (Naukowe, Warszawa-Poznan, 1977)
8. D. Pugh, Electric multipoles, polarizabilities and hyperpolarizabilities, in *Chemical Modelling: Applications and Theory*, vol. 1. The Royal Society of Chemistry, London, pp. 1–37 (2000)
9. W. Klemperer, V. Vaida, Molecular complexes in close and far away. Proc. Natl. Acad. Sci. U.S.A (PNAS) **103**(28), 10584–10588 (2006)
10. G. Maroulis, Interaction-induced electric properties, in *Chemical Modelling; Applications and Theory*, vol. 9, ed. by M. Springborg. The Royal Society of Chemistry, London, pp. 25–60 (2012)
11. J. Stone, *The Theory of Intermolecular Forces* (Clarendon Press, Oxford, 2002)
12. I.G. Kaplan, *Intermolecular Interactions: Physical Picture, Computational Methods and Model Potentials* (Wiley, Chichester, 2006)

Chapter 2
Theoretical Backgrounds
of Interaction-induced Theory

At present, a lot of fine manuals and books may be recommended to study the theoretical backgrounds of interaction of atoms and molecules [1–11]. However, in the main, the research and educational works are devoted to description of the energy of interacting molecules. At the same time, it is obvious that the knowledge of the electric and magnetic properties of interacting molecules [3–9] are also of interest. In accordance with the title of the book we focus on electric interactions as the strongest ones. Of course, the weak magnetic interactions are of interest for some specific problems of interacting molecules [2, 3], however, that is a special topic. In the short book it is not possible to describe in details the modern interaction-induced theory of molecules. Nevertheless, the general backgrounds of this theory are needed to be given to facilitate the reading of the book. In the next sections of this chapter some main definitions and formulas used in the book are given.

2.1 Multipole Electrical Moments

2.1.1 Coordinate System

In the book we will use for all molecular van der Waals complexes the Cartesian coordinate system (X, Y, Z) related to the complex (Fig. 2.1). The origin of this coordinate system (O_A) is usually placed on the molecule A. The vector $\boldsymbol{R}$ is directed from the point O_A of the molecule A to any point O_B of the molecule B. The relative orientation of molecules in a complex is determined by local coordinate systems (x_A, y_A, z_A) and (x_B, y_B, z_B) placed, accordingly, on the separate molecules A and B of the complex. The rotations of the molecules A and B are described by Euler angles, which determine the positions of the local coordinate systems (x_A, y_A, z_A) and (x_B, y_B, z_B) with respect to the complex coordinate system (X, Y, Z).

© The Author(s) 2017
V.N. Cherepanov et al., *Interaction-induced Electric Properties of van der Waals Complexes*,
SpringerBriefs in Electrical and Magnetic Properties of Atoms, Molecules, and Clusters,
DOI 10.1007/978-3-319-49032-8_2

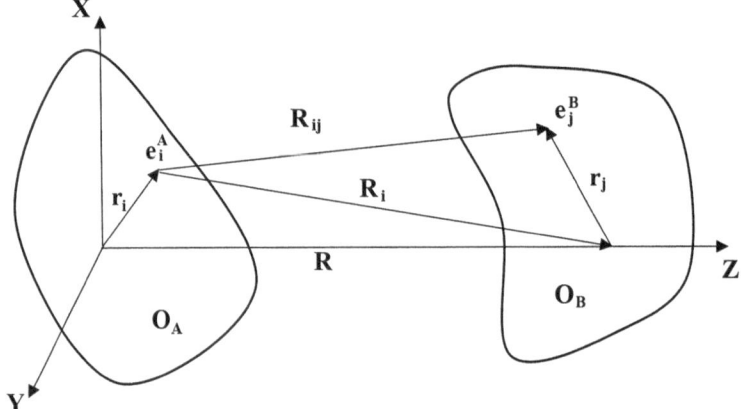

Fig. 2.1 Coordinate system of a molecular complex

2.1.2 Cartesian Definition

In this case the multipole electrical moments of the rank n for any molecule B can be defined in the form (see, for example, [1–6])

$$M_{\alpha\beta\ldots\nu}^{(n)(B)} = \frac{(-1)^n}{n!} \sum_j e_j^{(B)} r_j^{2n+1} \frac{\partial}{\partial r_{j\alpha}} \frac{\partial}{\partial r_{j\beta}} \cdots \frac{\partial}{\partial r_{j\nu}} \left(\frac{1}{r_j}\right), \qquad (2.1.1)$$

where the Greek indexes $\alpha, \beta, \ldots, \nu$ take the values X, Y, Z (the number of Greek indexes equals to n), e_j^B is any charge of the molecule B and $\mathbf{r}_j$ is the radius-vector of the charge relatively to the origin of local coordinate system O_B. The multipole electrical moments in the form (2.1.1) appear when the interaction Hamiltonian is expanded in the Taylor series about the point O_B (see Sect. 2.2). It is clear, that the well known simple moments of the lowest rank n such as the charge q, the electrical dipole and quadrupole moments (μ_α and $Q_{\alpha\beta}$) of the molecule B may be obtained from Eq. (2.1.1) if we put in it $n = 0$, 1 and 2 (analogously, for larger values of n the highest multipole moments can be written in the explicit form):

$$q^{(B)} = M^{(0)(B)} = \sum_j e_j^{(B)}, \qquad (2.1.2)$$

$$\mu_\alpha^{(B)} = M_\alpha^{(1)(B)} = \sum_j e_j^{(B)} r_{j\alpha},, \qquad (2.1.3)$$

$$\Theta_{\alpha\beta}^{(B)} = M_{\alpha\beta}^{(2)(B)} = \frac{1}{2} \sum_j e_j^{(B)} \left(3 r_{j\alpha} r_{j\beta} - r_j^2 \delta_{\alpha\beta}\right). \qquad (2.1.4)$$

The multipole tensors (2.1.1) have some important properties. One of them follows immediately from the Laplace equation $\Delta^2(1/r) = 0$, $M^{(n)}_{\alpha\alpha...v} = 0$ for any pair of equal suffixes (henceforth, we mean, as usually, the summation over the repeated indexes). From the definition of $M^{(n)}_{\alpha\beta...v}$ the permutation symmetry with respect to its suffixes occurs. Note that the total number of independent components of the multipole electrical moments of the rank n equals to $2n + 1$ for systems of C_1 symmetry. For the case of highly symmetric species the number of independent components is reduced.

2.1.3 Irreducible Spherical Tensor Definition

Sometimes it is useful to represent the multipole electrical moments in a spherical form. The spherical form of these moments allows us to apply effectively the theory of irreducible spherical tensor formalism. For this aim these 2^l-pole moments may be written in terms of the regular spherical harmonics using their both complex $R_{lm}(\mathbf{r})$ and real ($R_{lmc}(\mathbf{r})$ and $R_{lms}(\mathbf{r})$) forms defined, for $m > 0$, as

$$R_{lm}(\mathbf{r}) = \sqrt{\tfrac{4\pi}{2l+1}} r^l Y_{lm}(\theta, \varphi),$$
$$R_{lmc}(\mathbf{r}) = \sqrt{\tfrac{1}{2}}\left[(-1)^m R_{lm}(\mathbf{r}) + R_{l,-m}(\mathbf{r})\right], \qquad (2.1.5)$$
$$R_{lms}(\mathbf{r}) = -i\sqrt{\tfrac{1}{2}}\left[(-1)^m R_{lm}(\mathbf{r}) - R_{l,-m}(\mathbf{r})\right],$$

where r, θ and φ are the spherical coordinates of the vector argument $\mathbf{r}$.

Finally, the components of the spherical 2^l-pole electrical moments are written in the complex form as [1, 2]

$$Q_{lm} = \sum_j e_j R_{lm}(\mathbf{r}_j)$$

or in the real form

$$Q_{l\kappa} = \sum_j e_j R_{l\kappa}(\mathbf{r}_j),$$

where the labels $\kappa = mc, ms$ and take the values $0, 1c, 1s, 2c, 2s, ..., lc, ls$.

The linear relations between the Q_{lm} and $Q_{l\kappa}$ 2^l-pole electrical moments can be easily obtained from Eq. (2.1.5). It is clear from Eqs. (2.1.1) and (2.1.5) that the Cartesian components $M^{(n)}_{\alpha\beta...v}$ and the spherical components Q_{lm} are also in some linear relations.

2.1.4   Dependence on the Origin of the Coordinate System

In general case the multipole moments depend on the origin of the coordinate system (except for the charge q of a molecule). It follows from the definition (2.1.1) if the vector $\mathbf{r}$ is shifted on any vector $\boldsymbol{a}$. For uncharged molecules the origin dependence appears only for the multipole moments of the rank $n \geq 2$. As a result, because in the book only the uncharged molecules and complexes are considered, their dipole moments and connected with them polarizabilities and first hyperpolarizabilities do not depend on the origin of the coordinate system.

2.2   Interaction Hamiltonian

2.2.1   Large Separations

Let us consider briefly, following the works [1–6], how the Hamiltonian of two interacting molecules, which are far from each other, may be written to apply afterwards to the van der Waals complexes. It can be expected that for the far intermolecular distances the interaction energy of molecules is weak relative to the interaction energy within a molecule. As a result, the perturbation theory can be effectively used to describe the energy and other values of interacting molecules. So, the Hamiltonian of the molecular system becomes the sum of the Hamiltonian of the free molecules A and B $\left(H_0 = H_0^A + H_0^B \right)$ and the interaction Hamiltonian H':

$$H = H_0 + H', \tag{2.2.1}$$

where the interaction Hamiltonian H' restricted by the most strong Coulomb interactions has the form

$$H' = \sum_{ij} e_i^A e_j^B \left(R_{ij} \right)^{-1} = \sum_{j} e_j^B \varphi_j^B, \tag{2.2.2}$$

where $\varphi_j^B = \sum_i e_i^A \left(R_{ij} \right)^{-1}$ is the potential, caused by all charges of the molecule A, which effects on the jth charge of the molecule B. Then, if the potential φ_j^B is expanded in the Taylor series at the point O_B of the molecule B and the definition (2.1.1) is used, the following expression for the interaction Hamiltonian can be written in Coulomb form

$$H' = -\sum_{n} \frac{1}{(2n-1)!!} M_{\alpha\beta\ldots\nu}^{(n)B} F_{\alpha\beta\ldots\nu}^{B}. \tag{2.2.3}$$

Here, the functions $F^B_{\alpha\beta...\nu} = -(\nabla_\alpha\nabla_\beta...\nabla_\nu\varphi^B)_0$ are determined at the point O_B (hereafter, the Greek subscripts denote Cartesian components and the repeated subscripts imply summation over x, y, and z). In particular, $F^B_\alpha = -(\nabla_\alpha\varphi^B)_0$ is the electric field at the point O_B, $F^B_{\alpha\beta} = -(\nabla_\alpha\nabla_\beta\varphi^B)_0$ is the gradient of the electric field and so on. The potential

$$\varphi^B = \sum_i e^A_i (R_i)^{-1}$$

can also be expanded as

$$\varphi^B = \sum_i e^A_i \left[R^{-1} - r_{i\alpha}\nabla_\alpha R^{-1} + \frac{1}{2}r_{i\alpha}r_{i\beta}\nabla_\alpha\nabla_\beta R^{-1} - \cdots \right]$$
$$= \sum_n (-1)^n \frac{1}{(2n-1)!!} M^{(n)A}_{\alpha\beta...\nu} T^{(n)}_{\alpha\beta...\nu}. \tag{2.2.4}$$

Here the tensors $T^{(n)}_{\alpha\beta...\nu} = \nabla_\alpha\nabla_\beta...\nabla_\nu R^{-1}$ are proportional to $R^{(n+1)}$ (here n, as before, specifies the number of subscripts and is usually omitted hereinafter) and symmetric relative to the permutation for any pair of indexes. Also, as in the case of $M^{(n)}_{\alpha\alpha...\nu} = 0$, there is $T^{(n)}_{\alpha\alpha...\nu} = 0$. Note that the sign of the tensor $T^{(n)}_{\alpha\beta...\nu}$ depends on the definition of the direction of the vector $\boldsymbol{R}$ (from the atom A to the atom B or backward). As a result, $T^{A\to B}_{\alpha\beta...\nu} = (-1)^n T^{B\to A}_{\alpha\beta...\nu}$.

Then, the use of Eq. (2.2.4) leads to the expression for $F^{(B)}_{\alpha\beta...\nu}$ in the form

$$F^B_{\alpha\beta...\nu} = \sum_{n'} (-1)^{n'+1} \frac{1}{(2n'-1)!!} M^{(n')A}_{\alpha'\beta'...\nu'} T^{(n+n')}_{\alpha\beta...\nu\alpha'\beta'...\nu'}. \tag{2.2.5}$$

In the particular case of the electric field

$$F^B_\alpha = -T_\alpha q^A + T_{\alpha\beta}\mu^A_\beta - \frac{1}{3}T_{\alpha\beta\gamma}Q^A_{\beta\gamma} + \cdots \tag{2.2.6}$$

and for the gradient of the electric field

$$F^B_{\alpha\beta} = \nabla_\alpha F^B_\beta = -T_{\alpha\beta}q^A + T_{\alpha\beta\gamma}\mu^A_\gamma - \frac{1}{3}T_{\alpha\beta\gamma\delta}Q^A_{\gamma\delta} + \cdots. \tag{2.2.7}$$

Finally, the expression for the interaction Hamiltonian (2.2.3) can be written as

$$H' = \sum_{nn'} \frac{(-1)^{n'}}{(2n-1)!!(2n'-1)!!} M^{(n)B}_{\alpha\beta...\nu} M^{(n')A}_{\alpha'\beta'...\nu'} T^{(n+n')}_{\alpha\beta...\nu\alpha'\beta'...\nu'} \tag{2.2.8}$$

or in the expanded form

$$H' = Tq^A q^B + T_\alpha(q^A \mu_\alpha^B - q^B \mu_\alpha^A) + T_{\alpha\beta}(1/3q^A Q_{\beta\gamma}^B + 1/3q^B Q_{\beta\gamma}^A - \mu_\alpha^A \mu_\beta^B)$$
$$+ T_{\alpha\beta\gamma}(1/15q^A \Omega_{\alpha\beta\gamma}^B - 1/15q^B \Omega_{\alpha\beta\gamma}^A - 1/3\mu_\alpha^A Q_{\beta\gamma}^B + 1/3\mu_\alpha^B Q_{\beta\gamma}^A)$$
$$+ T_{\alpha\beta\gamma\delta}(1/9Q_{\alpha\beta}^A Q_{\gamma\delta}^B + \cdots) + \cdots. \tag{2.2.9}$$

It should be noted that in the presence of external electric field F_α^0 and its gradients $F_{\alpha\beta}^0$, $F_{\alpha\beta\gamma}^0$, ..., $F_{\alpha\beta...v}^0$ (in general case the external field may have some nonzero gradients) the total field is

$$F_{\alpha\beta...v} = F_{\alpha\beta...v}^0 + F_{\alpha\beta...v}^{(B)}. \tag{2.2.10}$$

As a consequence, the multipole moments of a molecule are also changed (see details in the Chaps. 3–5).Similarly, the expansion of R_{ij}^{-1} in Eq. (2.2.2) in terms of spherical harmonics allows us to define the interaction Hamiltonian H' through the spherical 2^l-pole electrical moments of interacting molecules A and B (see details, for instants, in [1, 2]):

$$H' = \sum_{l_1 l_2} \sum_{m=-l_<}^{l_<} \frac{F(l_1, l_2, m)}{R^{l_1 + l_2 + 1}} Q_{l_1 m}^{(A)} Q_{l_2, -m}^{(B)}, \tag{2.2.11}$$

where

$$F(l_1, l_2, m) = (-1)^{l_2} \frac{(l_1 + l_2)!}{[(l_1 + m)!(l_1 - m)!(l_2 + m)!(l_2 - m)!]^{1/2}} \tag{2.2.12}$$

and $l_< = \min\{l_1, l_2\}$.

2.2.2   Small Separations

It should be noted that for small intermolecular distances, when the electronic shells of interacting molecules begin to overlap, the effects of exchange interactions appear and the theoretical treatment discussed above can no more be applied for description of properties of interacting molecules. In this case, for small R, the interaction Hamiltonian H' in (2.2.2) can not be expanded in the Taylor series and the computational methods of quantum chemistry should be preferred. At present, due to the big progress in development of new methods and increasing of computational resources, computational chemistry is widely used. Among others, one should mark out ab inito methods of calculation. In general, ab initio methods give good qualitative and quantitative results and can provide better accuracy of calculations with decreasing of the size of system under investigation. The advantage of ab initio methods is that they do not use any empirical parameters. We can schematic distinguish the following law of increasing accuracy using a middle size basis set for single reference methods:

$$HF < MP2 \cong CCSD < SAPT \leq CCSD(T) \approx CCSD(T) - F12 < CCSDT\ldots,$$

where the conventional denotes are used for methods: HF—Hartree-Fock (or Self-Consistent Field (SCF)); MP2—second-order Møller-Plesset Perturbation theory; SAPT—Symmetry-Adapted Perturbation Theory; CCSD and CCSDT— Coupled Cluster theory with a full use of single (S), double (D) and triple (T) electron excitations; CCSD(T)—Coupled Cluster theory with a full use of single (S), double (D) and perturbative treatment of triple excitations. In 2007 a new explicitly correlated CCSD(T)-F12 approximation was presented and tested in Ref. [12]. This method proved to be more accurate and faster than conventional CCSD (T) method with the same basis set [12]. The considered methods (except HF) take into account the electron correlations which play an important role in van der Waals interactions, thus these methods could be successfully applied for solving problems of weakly-bound complexes. For example, CCSD(T) method allows to account for 99 % of correlation energy and at present it is one of the most popular methods applied to small molecules. However, it is difficult to apply this method for systems with a large number of particles due to big computational time and limited computational resources. For large molecular complexes the different variations of the density functional theory (DFT) or time dependent density functional theory (TDDFT) can be applied. The accuracy of these methods depends on the fortunate choose of exchange-correlation functionals. Recently, Wu et al. [13] have carried out the assessment of density functional methods in calculating the electric polarizability of 145 medium-size (3–14 atomic) organic molecules. They have found that at least three functionals among 34 considered should be used. They are devided into three groups: 1) one of M11 and M06-2X; 2) one of ωB97 and LC-τHTCH; 3) one of HISS and LC-ωPBE.

At last, it should be noted that a lot of semiempirical model potentials applied for both short and long separations are widely used now to solve different problems of physics and chemistry. A rather complete review of those one can find in the book of I. Kaplan [2].

2.3 Interaction Energy

2.3.1 Long-Range Distances

For the long-range intermolecular separations the perturbation theory can be applied. Within the perturbation theory using the interaction operator (2.2.3) one can obtain the energy of the pair of uncharged molecules A and B perturbed by a weak general static electric field as [5, 14]:

$$E^{AB} = E_A^0 + E_B^0 + \Delta E^{AB}, \tag{2.3.1}$$

where E_A^0 and E_B^0 are the unperturbed energies of interacting molecules and the interaction energy of the pair of molecules

$$\Delta E^{AB} = E_{electr}^{AB} + E_{ind}^{AB} + E_{disp}^{AB}. \tag{2.3.2}$$

Here the first-order perturbed energy is called the electrostatic energy E_{electr}^{AB}:

$$E_{electr}^{AB} = -\mu_\alpha^A F_\alpha^A - \frac{1}{3}\Theta_{\alpha\beta}^A F_{\alpha\beta}^A - \frac{1}{15}\Omega_{\alpha\beta\gamma}^A F_{\alpha\beta\gamma}^A - \frac{1}{105}\Phi_{\alpha\beta\gamma\delta}^A F_{\alpha\beta\gamma\delta}^A - \cdots. \tag{2.3.3}$$

The second-order perturbed energy is usually considered as two parts: induction (when only one interacting molecule changes its states) and dispersion (when both molecules change their states together).

The contribution to the induction energy E_{ind}^{AB} from two molecules can be written as $E_{ind}^{AB} = (1 + P^{AB})E_{ind}^A$, where P^{AB} permutes labels A and B and

$$
\begin{aligned}
E_{ind}^A = &-\frac{1}{2}\alpha_{\alpha\beta}^A F_\alpha^A F_\beta^A - \frac{1}{3}A_{\alpha,\beta\gamma}^A F_\alpha^A F_{\beta\gamma}^A - \frac{1}{6}C_{\alpha\beta,\gamma\delta}^A F_{\alpha\beta}^A F_{\gamma\delta}^A - \frac{1}{15}E_{\alpha,\beta\gamma\delta}^A F_\alpha^A F_{\beta\gamma\delta}^A \\
&-\frac{1}{105}D_{\alpha,\beta\gamma\delta\varepsilon}^A F_\alpha^A F_{\beta\gamma\delta\varepsilon}^A - \frac{1}{945}H_{\alpha,\beta\gamma\delta\varepsilon\phi}^A F_\alpha^A F_{\beta\gamma\delta\varepsilon\phi}^A - K_{\alpha\beta,\gamma\delta\varepsilon}^A F_{\alpha\beta}^A F_{\gamma\delta\varepsilon}^A - \cdots \\
&-\frac{1}{6}\beta_{\alpha\beta\gamma}^A F_\alpha^A F_\beta^A F_\gamma^A - \frac{1}{6}B_{\alpha\beta,\gamma\delta}^A F_\alpha^A F_\beta^A F_{\gamma\delta}^A - \frac{1}{30}M_{\alpha\beta,\gamma\delta\varepsilon}^A F_\alpha^A F_\beta^A F_{\gamma\delta\varepsilon}^A - \cdots \\
&-\frac{1}{24}\gamma_{\alpha\beta\gamma\delta}^A F_\alpha^A F_\beta^A F_\gamma^A F_\delta^A - N_{\alpha\beta\gamma,\delta\varepsilon}^A F_\alpha^A F_\beta^A F_\gamma^A F_{\delta\varepsilon}^A - \cdots.
\end{aligned} \tag{2.3.4}
$$

The dispersion energy E_{disp}^{AB} can be expressed in terms of the polarizabilities at imaginary frequency $i\omega$ [5, 11]:

$$
\begin{aligned}
E_{disp}^{AB} = &-\frac{1}{2\pi}T_{\alpha\beta}T_{\gamma\delta}\int_0^\infty \alpha_{\alpha\gamma}^A(i\omega)\alpha_{\beta\delta}^B(i\omega)d\omega \\
&-\frac{1}{3\pi}T_{\alpha\beta}T_{\gamma\delta\varepsilon}\int_0^\infty [\alpha_{\alpha\gamma}^A(i\omega)A_{\beta,\delta\varepsilon}^B(i\omega) - \alpha_{\alpha\gamma}^B(i\omega)A_{\beta,\delta\varepsilon}^A(i\omega)]d\omega \\
&-\frac{1}{6\pi}T_{\alpha\beta\gamma}T_{\delta\varepsilon\phi}\int_0^\infty [\alpha_{\alpha\delta}^A(i\omega)C_{\beta\gamma,\varepsilon\phi}^B(i\omega) + \alpha_{\alpha\delta}^B(i\omega)C_{\beta\gamma,\varepsilon\phi}^A(i\omega) \\
&-\frac{2}{3}A_{\beta,\varepsilon\phi}^B(i\omega)A_{\delta,\beta\gamma}^B(i\omega)]d\omega - \frac{1}{9\pi}T_{\alpha\beta}T_{\gamma\delta\varepsilon\phi}\int_0^\infty A_{\beta,\gamma\delta}^B(i\omega)A_{\beta,\varepsilon\phi}^B(i\omega)]d\omega + \cdots.
\end{aligned}
$$

$$\tag{2.3.5}$$

It is evident from (2.3.3)–(2.3.5) taking into account (2.2.5) and (2.2.10) that in the presence of an external field the multipole electric moments and polarizabilities of a molecule can be determined as the correspondent derivatives of the energy $E(F_\alpha, F_{\alpha\beta}, F_{\alpha\beta\gamma}, \ldots)$ with respect to an external field F_α^0 or field gradients $F_{\alpha\beta}^0, F_{\alpha\beta\gamma}^0, F_{\alpha\beta\gamma\delta}^0, \ldots$. In particular, these derivatives at $F_\alpha^0 = 0, F_{\alpha\beta}^0 = 0, F_{\alpha\beta\gamma}^0 = 0, \ldots$ give the values of multipole electric moments and polarizabilities of any molecule or van der Waals complex. For example,

$$\mu_\alpha = -\frac{\partial E}{\partial F_\alpha^0}\Bigg|_{F_\alpha^0=0,F_{\alpha\beta}^0=0,\ldots} \quad , \Theta_{\alpha\beta} = -3\frac{\partial E}{\partial F_{\alpha\beta}^0}\Bigg|_{F_\alpha^0=0,F_{\alpha\beta}^0=0,\ldots} \quad ,$$

$$\Omega_{\alpha\beta\gamma} = -15\frac{\partial E}{\partial F_{\alpha\beta\gamma}^0}\Bigg|_{F_\alpha^0=0,F_{\alpha\beta}^0=0,\ldots} \tag{2.3.6}$$

2.3.2   Small Distances

As mentioned in previous section for small distances between interacting molecules the methods of computational chemistry must be applied for calculation of any property of a supermolecule. It should be pointed the main sources of errors in ab initio calculations:

1. Basis set superposition error (it can be removed using the counterpoise (CP) correction scheme).
2. Error due to the incompleteness of the basis set (this error can be removed using the Complete Basis Set limit (CBS) extrapolation schemes).
3. Not full correlation.

It should be pointed out that ab initio methods are of high computational cost. They require a large amount of CPU time, disk storage and physical memory.

Basis Set Superposition Error (BSSE). The term BSSE was introduced by Liu and McLean in 1973 [15]. BSSE arises when two chemical fragments, A and B, approach each to another to form the AB supermolecule (or dimer) and the calculated interaction energy is unphysically overestimated. Note that A and B fragments can be both atoms and polyatomic species and present in every chemical bond. The conventional way to correct for BSSE is based on the Boys-Bernardi [16] counterpoise (CP) correction scheme. Using the CP correction one has to calculate the energy of a complex and energies of monomers in the basis of the whole complex for every geometrical arrangement. So, the CP-corrected interaction energy can be defined as

$$\Delta E^{CP} = E_{AB}(AB) - E_A(AB) - E_B(AB) \tag{2.3.7}$$

or the CP-corrected total potential energy of a complex

$$E^{CP} = E_{AB}(AB) + E_A(A) - E_A(AB) + E_B(B) - E_B(AB). \tag{2.3.8}$$

Here we use subscripts to denote the molecular species and the letters in parentheses refer to the (composite) basis used in the calculation. For example, $E_A(AB)$ is the energy of monomer A calculated using the basis set of the dimer *AB*. Of course, Eq. 2.3.8 can be generalized to the case of an arbitrary number of subsystems.

Basis Set Incompleteness Error (BSIE). It is known that in electronic structure calculations the basis sets are not complete. As a result, it is of interest to calculate the interaction energy in the Complete Basis Set (CBS) limit to eliminate BSIE.

Extrapolation schemes

If to use the CBS extrapolation for the BSSE-uncorrected energies, there is generally no monotonic convergence as is observed for BSSE-corrected ones [17]. But when we consider the energy values obtained successively, the convergence is rather systematic. Therefore, for the good convergence of energies to the CBS limit they have to use also rather large basis sets. For this purpose, the augmented correlation consistent aug-cc-pVXZ (X = 2 (D), 3 (T), 4 (Q), 5, etc., where X is a cardinal number of a basis set) basis sets (or, shortly, aVXZ) of Dunning [18] can be employed. At present, there are several CBS extrapolation schemes. The most known from them are the following schemes of Feller [19]

$$E_X^{HF} = E_{CBS}^{HF} + B\exp(-\alpha X), \tag{2.3.9}$$

of Truhlar [20]

$$\begin{aligned} E_X^{HF} &= E_{CBS}^{HF} + AX^{-\alpha}, \\ E_X^{corr} &= E_{CBS}^{corr} + BX^{-\beta}, \\ E_{CBS}^{tot} &= E_{CBS}^{HF} + E_{CBS}^{corr}, \end{aligned} \tag{2.3.10}$$

of Martin [21]

$$\begin{aligned} E_{CBS}^{tot} = {}&(X+3/2)^4/[(X+3/2)^4 - (X+1/2)^4]E_{X+1}^{tot} \\ &- (X+1/2)^4/[(X+3/2)^4 - (X+1/2)^4]E_X^{tot} \end{aligned} \tag{2.3.11}$$

and of Helgaker [22]

$$\begin{aligned} E_X^{HF} &= E_{CBS}^{HF} + B\exp(-\alpha X), \\ E_X^{corr} &= E_{CBS}^{corr} + AX^{-3}, \\ E_{CBS}^{tot} &= E_{CBS}^{HF} + E_{CBS}^{corr}. \end{aligned} \tag{2.3.12}$$

Here E_X^{HF}, E_X^{corr} and E_X^{tot} are the Hartree-Fock, correlation and total energy, accordingly; B, A and α are the parameters to be optimized. The subscripts "X" and "CBS" correspond to the energy calculated using aug-cc-pVXZ basis set and the energy obtained in the complete basis set limit. It should be noted that Helgaker's extrapolation scheme is a three-point one for the Hartree-Fock energy and a two-point for the correlation energy.

It should be pointed out that such extrapolation schemes to the CBS limit can be also applied for the accurate calculations of bond lengths and angles [17].

Bond functions

There is an other way to get results closer to those obtained in the CBS limit: placing additional bond functions between molecules on a ghost atom. It will speed up the convergence of calculations to the CBS limit. There are two ways of placing the functions: first one, and more often used, at mid distance on the vector R connecting two molecules/atoms; the second one, at the distance on vector R that corresponds to the center of mass of the complex.

2.4 (Hyper)Polarizabilities

When any molecule is in the external electric field F_α^0 the new electric property of it, connected with the polarization of the molecule, appears. The polarization leads to the dependence of the electrical dipole moment on the field F_α^0. This dependence may be formally represented by the Taylor series (see 2.3.1 and 2.3.2)

$$\mu_\alpha = \mu_\alpha^0 + \alpha_{\alpha\beta}F_\beta^0 + \frac{1}{2!}\beta_{\alpha\beta\gamma}F_\beta^0 F_\gamma^0 + \frac{1}{3!}\gamma_{\alpha\beta\gamma\delta}F_\beta^0 F_\gamma^0 F_\delta^0 + \cdots, \qquad (2.4.1)$$

where the tensors $\alpha_{\alpha\beta}$, $\beta_{\alpha\beta\gamma}$ and $\gamma_{\alpha\beta\gamma\delta}$ are the polarizability, first hyperpolarizability and second hyperpolarizability tensors of a system, respectively. It is seen from Eq. (2.4.1) that the (hyper)polarizability tensors of a system can be defined as

$$\alpha_{\alpha\beta} = \left.\frac{\partial\mu_\alpha}{\partial F_\beta^0}\right|_{F^0=0}, \quad \beta_{\alpha\beta\gamma} = \left.\frac{\partial\mu_\alpha}{\partial F_\beta^0 F_\gamma^0}\right|_{F^0=0}, \quad \gamma_{\alpha\beta\gamma\delta} = \left.\frac{\partial\mu_\alpha}{\partial F_\beta^0 F_\gamma^0 F_\delta^0}\right|_{F^0=0}, \qquad (2.4.2)$$

or as the corresponding derivatives of the energy (see 2.3.3–2.3.5). It is also clear how other higher hyperpolarizabilities of a system can be determined. Such procedure is usually used in the computational codes.

When the external field is not strong and changes over time, the time-dependent perturbation theory can be applied to obtain the analytical expression for the (hyper) polarizabilities. For this, it is sufficient to consider the dipole approximation when the interaction operator $H'(t) = -\mu_\alpha F_\alpha^0(t)$ where $F_\alpha^0(t) = F_\alpha^0 \exp(-i\omega t)$ is any harmonic field. Then the average dipole moment of a volume V is determined as

$$\bar{\mu}_\alpha(t) = Tr(\rho(t), \mu_\alpha) = Tr(\tilde{\rho}(t), \tilde{\mu}_\alpha) \tag{2.4.3}$$

where $\rho(t)$ and $\tilde{\rho}(t)$ are the density matrices (the tilde sign over the operators denotes, as usual, that these operators are written in the interaction representation: $\tilde{A} = \exp\left(\frac{i}{\hbar}H_0t\right)A\exp\left(-\frac{i}{\hbar}H_0t\right)$) and the following equation for $\tilde{\rho}(t)$ is fulfilled

$$i\hbar\frac{\partial\tilde{\rho}}{\partial t} = [\tilde{H}'(t), \tilde{\rho}]. \tag{2.4.4}$$

Assuming that $\tilde{\rho}(t)$ can be expanded in a series in powers of the perturbation

$$\tilde{\rho}(t) = \sum_k \tilde{\rho}^{(k)}(t), \tag{2.4.5}$$

the Eq. (2.4.3) gives the iterative equation

$$i\hbar\frac{\partial\tilde{\rho}^{(k)}}{\partial t} = \left[\tilde{H}'(t), \tilde{\rho}^{(k-1)}\right], \tag{2.4.6}$$

where in the first approximation

$$\tilde{\rho}^{(1)}(t) = \frac{1}{i\hbar}\int\limits_{-\infty}^{t} [\tilde{H}'(t'), \rho_0]dt' \tag{2.4.7}$$

and ρ_0 is the equilibrium density matrix in energetic representation when external fields are absencet (usually, it is the Gibbs canonical distribution).

In this way, for the equilibrium density matrix ρ_0 the susceptibility tensors $\chi_{\alpha\beta\gamma...}^{(k)}$ can be determined for any kth approximation. These tensors are determined through the $\alpha_{\alpha\beta}$, $\beta_{\alpha\beta\gamma}$, $\gamma_{\alpha\beta\gamma\delta}$ polarizabilities and so on for $k = 1, 2, 3,...$ respectively as:

$$\chi_{\alpha\beta\gamma...}^{(k)} = \frac{1}{V}\sum_n e^{\frac{F-E_n}{kT}}\left\langle n\left|\hat{\chi}_{\alpha\beta\gamma...}^{(k)}\right|n\right\rangle, \tag{2.4.8}$$

where F is the Helmholtz free energy and $\langle n|\hat{\chi}_{\alpha\beta}^{(1)}|n\rangle \equiv \alpha_{\alpha\beta}^{(n)}$, $\langle n|\hat{\chi}_{\alpha\beta\gamma}^{(2)}|n\rangle \equiv \beta_{\alpha\beta\gamma}^{(n)}$, $\langle n|\hat{\chi}_{\alpha\beta\gamma\delta}^{(3)}|n\rangle \equiv \gamma_{\alpha\beta\gamma\delta}^{(n)}$ and so on are, respectively, the polarizability, first hyperpolarizability, second hyperpolarizability and higher hyperpolarizability tensors of a molecule being at the state n (thereafter, for the ground state the index n for polarizabilities will be omitted). It is clear also from Eqs. (2.4.2) and (2.4.3) that any (hyper)polarizability tensor depends on the frequencies of external fields (see the details in [1, 3–5, 11]).

References

1. A.J. Stone, *The Theory of Intermolecular Forces* (Clarendon Press, Oxford, 2002)
2. I.G. Kaplan, *Intermolecular Interactions: Physical Picture, Computational Methods and Model Potentials* (Wiley, New Jersey, 2006)
3. A. Salam, *Molecular Quantum Electrodynamics: Long-Range Intermolecular Interactions* (Wiley, New Jersey, 2010)
4. S. Kielich, *Molekularna Optyka Nieliniowa (Nonlinear Molecular Optics)* (Naukowe, Warszawa-Poznan, 1977)
5. A.D. Buckingham, in *Intermolecular Interaction: From Diatomic to Biopolymers*, ed. by B. Pullman (Wiley, New York, 1978)
6. G. Maroulis (ed.), *Atoms, Molecules and Clusters in Electric Fields. Theoretical Approaches to the Calculation of Electric Polarizability* (Imperial College Press, Singapore, 2006)
7. G. Maroulis (ed.), *Computational Aspects of Electric Polarizability Calculations: Atoms, Molecules and Clusters* (IOS Press, Amsterdam, 2006)
8. G. Maroulis, T. Bancewicz, B. Champagne and A.D. Buckingham (eds.), *Atomic and Molecular Nonlinear Optics: Theory, Experiment and Computation. A Homage to the Pioneering Work of Stanislaw Kielich (1925–1993)* (IOS Press Inc., Amsterdam, Netherland, 2011)
9. G. Birnbaum (ed.), *Phenomena Induced by Intermolecular Interactions* (Plenum, New York, 1985)
10. G.C. Tabitz, M.N. Neuman (eds.), *Collision- and Interaction-Induced Spectroscopy* (Kluwer, Dordrecht, 1995)
11. A.D. Buckingham, Permanent and Induced Molecular Moments and Long-Range Intermolecular Forces. Adv. Chem. Phys. **12**, 107–142 (1967)
12. T.B. Adlerc, G. Knizia, H.-J. Werner, A simple and efficient CCSD(T)-F12 approximation. J. Chem. Phys. **127**(22), 221106 (2007)
13. T. Wu, Y.N. Kalugina, A.J. Thakkar, Choosing a density functional for static molecular polarizabilities. Chem. Phys. Lett. **635**, 257 (2015)
14. B. Linder, R.A. Kromhout, Van der Waals induced dipoles. J. Chem. Phys. **84**(5), 2753–2760 (1986)
15. B. Liu, A.D. McLean, Accurate calculation of the attractive interaction of two ground state helium atoms. J. Chem. Phys. **59**(8), 4557–4558 (1973)
16. S.F. Boys, F. Bernardi, The calculations of small molecular interaction by the difference of separate total energies. Some procedures with reduced error. Mol. Phys. **19**(4), 553–566 (1970)
17. B. Paizs, P. Salvador, A.G. Csaszar, M. Duran, S. Suhai, Intermolecular bond lengths: extrapolation to the basis set limit on uncorrected and BSSE-corrected potential energy hypersurfaces. J. Comp. Chem. **22**(2), 196–207 (2001)
18. T.H. Dunning Jr., Gaussian basis sets for use in correlated molecular calculations. I. The atoms boron through neon and hydrogen. J. Chem. Phys. **90**(2), 1007–1023 (1989)
19. D. Feller, Application of systematic sequences of wave functions to the water dimer. J. Chem. Phys. **96**(8), 6104–6114 (1992)
20. D.G. Truhlar, Basis-set extrapolation. Chem. Phys. Lett. **294**(1–3), 45–48 (1998)
21. J.M.L. Martin, Ab initio total atomization energies of small molecules—towards the basis set limit. Chem. Phys. Lett. **259**(5–6), 669–678 (1996)
22. A. Halkier, T. Helgaker, P. Jnrgensen, W. Klopper, J. Olsen, Basis-set convergence of the energy in molecular Hartree-Fock calculations. Chem. Phys. Lett. **302**(5–6), 437–446 (1999)

Chapter 3
Interaction-induced Dipole Moment

3.1 General Backgrounds

3.1.1 Computational Features

In order to evaluate the dipole moment, the finite-field method [see Eq. (2.3.2)] described by Cohen and Roothaan [1] is often employed. This approach is now implemented practically in all computational codes (Gaussian [2], Molpro [3] and others). One can propose several techniques to obtain the dipole moment in this way. The first technique is evident following the definition (2.3.2). In this case the dipole moment components can be determined easily as the first derivatives of the energy $E(F_\alpha^0)$ with respect to the external field F_α^0 using the most simple 2-point expression [with errors of order $\left(F_\alpha^0\right)^2$]:

$$\mu_\alpha = -\frac{E(F_\alpha^0) - E(-F_\alpha^0)}{2F_\alpha^0} \qquad (3.1.1)$$

or using the more correct 5-point stencil formula (with errors of order $\left(F_\alpha^0\right)^4$). However, in this way, the higher polarizabilities give some contributions to μ_α. And to decrease these contributions the values of fields used have to be chosen very accurately. To remove the contributions of higher polarizabilities it is enough to write the system of equations for $E(F_\alpha^0)$ in Eq. (2.3.1)–(2.3.4) at the values of $\pm F_\alpha^0, \pm 2F_\alpha^0, \pm 4F_\alpha^0, \ldots$ for the case of homogeneous field. Then, solving this system with respect to μ_α, the dependence of the dipole moment on arbitrary higher polarizabilities can be removed. This procedure can be also applied to the case of (hyper)polarizabilities. The generalization for the nonhomogeneous external electric field can be applied for the multipole moments and high-order polarizabilities of any molecule. In this way, Maroulis [4] has proposed the more accurate formula restricted by the term in (2.3.4) including the second hyperpolarizability:

© The Author(s) 2017
V.N. Cherepanov et al., *Interaction-induced Electric Properties of van der Waals Complexes*,
SpringerBriefs in Electrical and Magnetic Properties of Atoms, Molecules, and Clusters,
DOI 10.1007/978-3-319-49032-8_3

$$\mu_\alpha = \frac{256 D_\alpha(F_\alpha^0) - 40 D_\alpha(2F_\alpha^0) + D_\alpha(4F_\alpha^0)}{180 F_\alpha^0}, \tag{3.1.2}$$

where $D_\alpha(F_\alpha^0) = -\frac{E(-F_\alpha^0) - E(F_\alpha^0)}{2}$.

3.1.2   Long-Range Analytical Formalism. Induction and Dispersion Contributions

To obtain the induced dipole moment of the molecule A being in the weak electric field of the molecule B one should calculate the first derivative of the energy E^{AB} with respect to the external field F_α^0. The electrostatic interactions (Eq. 2.3.3) give the permanent dipole moment $\mu_\alpha^{(0)AB} = \mu_\alpha^{(0)A} + \mu_\alpha^{(0)B}$, the induction interactions (Eq. 2.3.4) give the contribution to the induction dipole moment μ_{ind}^{AB} and the dispersion interactions (Eq. 2.3.5) give the contribution to the dispersion dipole moment μ_{disp}^{AB}. Therefore, the total dipole moment μ_α^{AB} is the sum of the permanent, induction and dispersion dipole moments:

$$\mu_\alpha^{AB} = \mu_\alpha^A + \mu_\alpha^B = (1 + P^{AB})\mu_\alpha^A, \tag{3.1.3}$$

where $\mu_\alpha^A = \mu_\alpha^{(0)A} + \mu_\alpha^{ind,A} + \mu_\alpha^{disp,A}$. The induction part of the dipole moment of the molecule A has the form (see Eq. 2.3.6)

$$\mu_\alpha^{ind,A} = -\frac{\partial E_{ind}^A}{\partial F_\alpha^0} = \alpha_{\alpha\beta}^A F_\beta^A + \frac{1}{3} A_{\alpha,\beta\gamma}^A F_{\beta\gamma}^A + \frac{1}{15} E_{\alpha,\beta\gamma\delta}^A F_{\beta\gamma\delta}^A + \frac{1}{105} D_{\alpha,\beta\gamma\delta\varepsilon}^A F_{\beta\gamma\delta\varepsilon}^A$$

$$+ \frac{1}{945} H_{\alpha,\beta\gamma\delta\varepsilon\varphi}^A F_{\beta\gamma\delta\varepsilon\varphi}^A + \cdots + \frac{1}{2} \beta_{\alpha\beta\gamma}^A F_\beta^A F_\gamma^A + \frac{1}{3} B_{\alpha\beta,\gamma\delta}^A F_\beta^A F_{\gamma\delta}^A + \frac{1}{15} M_{\alpha\beta,\gamma\delta\varepsilon}^A F_\beta^A F_{\gamma\delta\varepsilon}^A + \cdots$$

$$+ \frac{1}{6} \gamma_{\alpha\beta\gamma\delta}^A F_\beta^A F_\gamma^A F_\delta^A + 3 N_{\alpha\beta\gamma,\delta\varepsilon}^A F_\beta^A F_\gamma^A F_{\delta\varepsilon}^A. \cdots$$

$$\tag{3.1.4}$$

For further consideration of the dipole moment, polarizability and hyperpolarizability of a complex we need also some expressions for multipole moments (see 2.3.1–2.3.6):

2^2-pole (quadrupole) moment

$$\Theta_{\alpha\beta}^A = \Theta_{\alpha\beta}^{(0)A} + A_{\gamma,\delta\varepsilon}^A F_\gamma^A + \frac{1}{2} B_{\gamma,\delta,\alpha\beta}^A F_\gamma^A F_\delta^A + C_{\alpha\beta,\gamma\delta}^A F_{\gamma\delta}^A + \cdots, \tag{3.1.5}$$

2^3-pole (octupole) moment

$$\Omega^A_{\alpha\beta\gamma} = \Omega^{(0)A}_{\alpha\beta\gamma} + E^A_{\delta,\alpha\beta\gamma}F^A_{\delta} + \cdots, \tag{3.1.6}$$

2^4-pole (hexadecapole) moment

$$\Phi^A_{\alpha\beta\gamma\delta} = \Phi^{(0)A}_{\alpha\beta\gamma\delta} + D^A_{\varepsilon,\alpha\beta\gamma\delta}F^A_{\varepsilon} + \cdots, \tag{3.1.7}$$

and 2^5-pole moment

$$\Xi^A_{\alpha\beta\gamma\delta\varepsilon} = \Xi^{(0)A}_{\alpha\beta\gamma\delta\varepsilon} + H^A_{\varphi,\alpha\beta\gamma\delta\varepsilon}F^A_{\varphi} + \cdots. \tag{3.1.8}$$

If the other highest multipole moments are needed the procedure to obtain their expressions is evident. So, in order to obtain the induced dipole moment of the molecule A, we expand $\mu^{ind,A}_{\alpha}$ in (3.1.4) using equations for the field and field gradients (Eq. 2.2.5). Then Eqs. (3.1.4)–(3.1.18) for the dipole, quadrupole and other moments should be also employed. And assuming $F^0_{\alpha} = 0$ for the expressions obtained, the induction dipole moment of the molecule A can be written in the general case for neutral systems as (see also [5])

$$\begin{aligned}
\mu^{ind,A}_{\alpha} &= \alpha^A_{\alpha\beta}\mu^B_{\gamma}T_{\beta\gamma} + \frac{1}{3}\alpha^A_{\alpha\beta}\Theta^B_{\gamma\delta}T_{\beta\gamma\delta} - \frac{1}{3}A^A_{\alpha,\beta\gamma}\mu^B_{\delta}T_{\beta\gamma\delta} - \frac{1}{9}A^A_{\alpha,\beta\gamma}\Theta^B_{\delta\varepsilon}T_{\beta\gamma\delta\varepsilon} \\
&+ \frac{1}{15}\alpha^A_{\alpha\beta}\Omega^B_{\gamma\delta\varepsilon}T_{\beta\gamma\delta\varepsilon} + \frac{1}{15}E^A_{\alpha,\beta\gamma\delta}\mu^B_{\varepsilon}T_{\beta\gamma\delta\varepsilon} + \frac{1}{105}\alpha^A_{\alpha\beta}\Phi^B_{\gamma\delta\varepsilon\varphi}T_{\beta\gamma\delta\varepsilon\varphi} - \frac{1}{105}D^A_{\alpha,\beta\gamma\delta\varepsilon}\mu^B_{\varphi}T_{\beta\gamma\delta\varepsilon\varphi} \\
&- \frac{1}{45}A^A_{\alpha,\beta\gamma}\Omega^B_{\delta\varepsilon\varphi}T_{\beta\gamma\delta\varepsilon\varphi} + \frac{1}{45}E^A_{\alpha,\beta\gamma\delta}\Theta^B_{\varepsilon\varphi}T_{\beta\gamma\delta\varepsilon\varphi} + \frac{1}{945}\alpha^A_{\alpha\beta}\Xi^B_{\gamma\delta\varepsilon\varphi\nu}T_{\beta\gamma\delta\varepsilon\varphi\nu} - \frac{1}{315}A^A_{\alpha,\beta\gamma}\Phi^B_{\delta\varepsilon\varphi\nu}T_{\beta\gamma\delta\varepsilon\varphi\nu} \\
&- \frac{1}{315}D^A_{\alpha,\beta\gamma\delta\varepsilon}\Theta^B_{\varphi\nu}T_{\beta\gamma\delta\varepsilon\varphi\nu} + \frac{1}{225}E^A_{\alpha,\beta\gamma\delta}\Omega^B_{\varepsilon\varphi\nu}T_{\beta\gamma\delta\varepsilon\varphi\nu} + \frac{1}{945}H^A_{\alpha,\beta\gamma\delta\varepsilon\varphi}\mu^B_{\nu}T_{\beta\gamma\delta\varepsilon\varphi\nu} + \frac{1}{2}\beta^A_{\alpha\beta\delta}\mu^B_{\gamma}\mu^B_{\delta}T_{\beta\gamma}T_{\delta\varepsilon} \\
&+ \alpha^A_{\alpha\beta}\mu^A_{\gamma}\alpha^B_{\delta\varepsilon}T_{\beta\delta}T_{\gamma\varepsilon} + \frac{2}{3}\alpha^A_{\alpha\beta}\mu^A_{\delta}A^B_{\delta,\varepsilon\varphi}T_{\beta\varphi}T_{\gamma\delta\varepsilon} - \frac{1}{3}\alpha^A_{\alpha\beta}\Theta^A_{\gamma\delta}\alpha^B_{\varepsilon\varphi}T_{\beta\varepsilon}T_{\varphi\gamma\delta} - \frac{1}{3}A^A_{\alpha,\beta\gamma}\mu^A_{\delta}\alpha^B_{\varepsilon\varphi}T_{\gamma\varepsilon}T_{\varphi\beta\delta} \\
&+ \frac{1}{3}\beta^A_{\alpha\beta\delta}\mu^B_{\delta}\Theta^B_{\varepsilon\varphi}T_{\beta\delta}T_{\gamma\varepsilon\varphi} - \frac{1}{3}B^A_{\alpha,\beta\gamma\delta}\mu^B_{\varepsilon}\mu^B_{\varphi}T_{\beta\varepsilon}T_{\gamma\delta\varphi} + \cdots.
\end{aligned} \tag{3.1.9}$$

For simplicity, hereinafter for final formulas the upper index (0) for all operators is omitted. Accordingly, using Eq. (2.3.5) the accurate dispersion contribution through the order R^{-7} to the dipole is written as:

$$\mu^{disp,AB}_{\alpha} = -\frac{\partial E^{AB}_{disp}}{\partial F^0_{\alpha}}\Bigg|_{F^0_{\alpha}=0} = \mu^{disp,A}_{\alpha} + \mu^{disp,B}_{\alpha},$$

where

$$\begin{aligned}
\mu^{disp,A}_{\alpha} &= \frac{1}{2\pi}T_{\beta\delta}T_{\gamma\varepsilon}\int_0^{\infty}d\omega\,\beta^A_{\alpha\beta\gamma}(0,i\omega,-i\omega)\alpha^B_{\delta\varepsilon}(i\omega) \\
&+ \frac{1}{3\pi}T_{\beta\varepsilon}T_{\gamma\delta\varphi}\int_0^{\infty}d\omega\,B^A_{\alpha,\varepsilon,\delta\varphi}(0,i\omega,-i\omega)\alpha^B_{\beta\gamma}(i\omega) - \frac{1}{3\pi}T_{\beta\varepsilon}T_{\gamma\delta\varphi}\int_0^{\infty}d\omega\,\beta^A_{\alpha\beta\gamma}(0,i\omega,-i\omega)A^B_{\varepsilon,\delta\varphi}(i\omega).
\end{aligned} \tag{3.1.10}$$

It should be noted that it is not always easy to calculate the dynamic highest (hyper)polarizabilities at imaginary frequency $i\omega$. However, some good approximations can be used to estimate their values [6–9]. These approximations are based on the "constant ratio" approximations (CRA) that allows us to express the dispersion contribution to the dipole moment in terms of static (hyper)polarizabilities. Let us consider, for example, the more correct approximation CRA2 from them proposed in the original paper [9]. This approximation gives the simple relations for the integrals appeared in Eq. (3.1.10). Consider for illustration, in detail, first of them:

$$\int_{0}^{\infty} d\omega \beta_{\alpha\beta\gamma}^{A,B}(0, i\omega, -i\omega)\alpha_{\delta\varepsilon}^{B,A}(i\omega) = \frac{\pi}{3}\frac{I_{\beta\alpha}^{A,B}}{I_{\alpha\alpha}}C_6 \qquad (3.1.11)$$

where the following notations are used

$$C_6 = \frac{3}{\pi}\int_{0}^{\infty} d\omega \alpha^{A}(i\omega)\alpha^{B}(i\omega) \qquad (3.1.12)$$

and the "constant ratio"

$$\frac{I_{\beta\alpha}^{A,B}}{I_{\alpha\alpha}} = \frac{\int_{0}^{\infty} d\omega \beta_{\alpha\beta\gamma}^{A,B}(0, i\omega, -i\omega)\alpha_{\delta\varepsilon}^{B,A}(i\omega)}{\int_{0}^{\infty} d\omega \alpha^{A}(i\omega)\alpha^{B}(i\omega)} \cdot \qquad (3.1.13)$$

Here, the $\alpha^{A,B}$ is the mean polarizability of a molecule A(or B):

$$\alpha^{A,B} = \frac{1}{3}(\alpha_{xx}^{A,B} + \alpha_{yy}^{A,B} + \alpha_{zz}^{A,B}). \qquad (3.1.14)$$

The last integral can be estimated rather well using Unsöld approximation [10] that gives the following expressions for the (hyper)polarizabilities:

$$\alpha_{\alpha\beta}(i\omega) = \alpha_{\alpha\beta}(0)\frac{\Omega^2}{\Omega^2 + \omega^2},$$
$$\beta_{\alpha\beta\gamma}(0, i\omega, -i\omega) = \beta_{\alpha\beta\gamma}(0, 0, 0)\frac{\Omega^2}{3}\frac{3\Omega^2 + \omega^2}{(\Omega^2 + \omega^2)^2}. \qquad (3.1.15)$$

Here Ω is any average excitation frequency for the molecule. Therefore, after taking the integrals over ω the following estimation of (3.1.13) can be obtained

$$I^{A,B}_{\beta\alpha} \cong \frac{\beta^{A,B}_{\alpha\beta\gamma}(0,0,0)\alpha^{B,A}_{\delta\varepsilon}(0)}{\alpha^A(0)\alpha^B(0)}\left[\frac{1+\frac{2}{3}\Delta_{1,2}}{1+\Delta_{1,2}}\right], \qquad (3.1.16)$$

where

$$\Delta_1 = \frac{\Omega_B}{\Omega_A}, \Delta_2 = \frac{\Omega_A}{\Omega_B}.$$

Analogously, one can get the expressions for other integrals in Eq. (3.1.10). As a result, assuming that $\Omega_A \cong \Omega_B$, the expression for $\mu^{disp,A}_\alpha$ takes the form

$$\mu^{disp,AB}_\alpha = \left[\beta^A_{\alpha\beta\gamma}\alpha^B_{\delta\varepsilon} + \beta^B_{\alpha\beta\gamma}\alpha^A_{\delta\varepsilon}\right]\frac{5T_{\beta\delta}T_{\gamma\varepsilon}C_6}{36\alpha^A\alpha^B}$$
$$+ \left[B^A_{\alpha,\beta,\delta\varphi}\alpha^B_{\beta\gamma} - B^B_{\alpha,\beta,\delta\varphi}\alpha^A_{\beta\gamma} - \beta^A_{\alpha\beta\gamma}A^B_{\varepsilon,\delta\varphi} + \beta^B_{\alpha\beta\gamma}A^B_{\varepsilon,\delta\varphi}\right]\frac{5T_{\beta\varepsilon}T_{\gamma\delta\varphi}C_6}{54\alpha^A\alpha^B}.$$
$$(3.1.17)$$

Thus, having necessary properties of free molecules, it is not difficult to calculate the interaction-induced dipole moment using suggested formulas. As the components of the properties (polarizabilities, multipole moments, etc.) are dependent, in general case, on the orientation of molecules (Appendix A) we have a multidimensional surface of the dipole moment for a complex.

3.1.3 Exchange Interaction Contributions

When the valence shells of interacting species are weakly overlapped, the analytical formalism can be applied to describe their electrical properties taking into account exchange interactions. In this case, to take into account the exchange effects, the asymptotic methods [11, 12] could be used. These methods may only be applied in a range of R where a weak overlapping of the valence electron shells of interacting systems takes places. Such situation is typical for the ranges of R corresponding to potential wells of van der Waals complexes. Let us consider a case of two interacting atoms with the valence s-electrons. In this case, the exchange interaction of atoms can be approximately considered as an exchange interaction of two valence electrons (by one from each atom). Then the two-electron (one electron from atom A and one from atom B) molecular wave function of the state n can be written in the form:

$$\Psi_n(r_1, r_2, R) = c^{(1)}_n\psi^{(1)}_n(r_1, r_2, R) + c^{(2)}_n\psi^{(2)}_n(r_1, r_2, R), \qquad (3.1.18)$$

where

$$\psi_n^{(1)}(r_1, r_2, R) = \left[\varphi^{(A)}(r_1, R)\varphi^{(B)}(r_2, R)\chi_I(r_1, r_2, R)\right]_n,$$
$$\psi_n^{(2)}(r_1, r_2, R) = \left[\varphi^{(A)}(r_2, R)\varphi^{(B)}(r_1, R)\chi_{II}(r_1, r_2, R)\right]_n. \tag{3.1.19}$$

Here $\varphi^{(A)}(r_1, R)$, $\varphi^{(B)}(r_1, R)$ and $\varphi^{(A)}(r_2, R)$, $\varphi^{(B)}(r_2, R)$, are asymptotic wave functions of the first and of the second electrons located near corresponding atom cores. Equations (3.1.18) and (3.1.19) are written in the molecular coordinate system in which the interacting atoms are located on the axis z, and the center of the interatomic separation is taken as the origin of coordinates. In this coordinate system r_1 and r_2 are the coordinates of the first and of the second electrons. The functions $\chi_I(r_1, r_2, R)$ and $\chi_{II}(r_1, r_2, R)$ accounting the interaction of electrons with each other and with extraneous nuclei have complicated forms and are given in [12].

The function $\varphi(r, R)$ can be obtained from the asymptotic radial wave function of a valence electron of a neutral atom. This radial wave in the coordinate system with the origin in the atom nuclear has the form [11]:

$$\varphi(r) = A_0 r^{1/\beta - 1} \exp(-r\beta), \tag{3.1.20}$$

where $\beta^2/2$ is the atom ionization potential and the value of the asymptotic coefficient A_0 depends on the electron distribution in the internal zone of the atom. The function $\varphi(r, R)$ is obtained from the function $\varphi(r)$ by transition from the atomic coordinate system to the coordinate system of the complex.

Then the exchange interaction contribution into α-component of the dipole moment for two interacting atoms may be represented as [13]

$$\mu_\alpha^{exch} = \left\langle \psi_n^{(1)}(r_1, r_2, R) \left| \mu_\alpha \right| \psi_n^{(2)}(r_1, r_2, R) \right\rangle^{exch}$$
$$= B_\alpha(\chi_A, \theta_A, \varphi_a, \chi_B, \theta_B, \varphi_B) R^\delta \exp(-\eta R) \tag{3.1.21}$$

where the parameters δ and η are determined by the ionization potentials of the atoms A and B. at the electronic state n. For small molecules we can conserve the form of R-dependence of the exchange dipole moment components like the form for interacting atoms in Eq. (3.1.21). This form is supposed to be the same for all components and doesn't depend on the mutual orientation of molecules in the complex. The orientational dependence of μ_α^{exch} (through the Euler angles $\chi_A, \theta_A, \varphi_a, \chi_B, \theta_B, \varphi_B$) is introduced by B_α parameter which is weakly depended on R for the region of small overlapping of electron shells. It should be noted, that the size of interacting molecules is accounted only by the B_α parameter.

The results obtained for atoms with the valence s-electrons can also be applied, after minimal changes, to interacting atoms having the valence electrons of nonzero orbital moment l. Indeed, the exchange interaction occurs in the range of electron coordinates near the axis z where the angular wave functions of the electrons vary slightly and so may be substituted by their values on the z axis. Therefore, for this case taking into account the exchange interaction is reduced to the problem considered above with the only difference that for the case of electrons with nonzero orbital momentum the coefficient A_0 in Eq. (3.1.20) should be multiplied by $\sqrt{2l+1}$ [12].

3.2   Dipole Moment of van der Waals Complexes

For ab initio calculations of all properties of van der Waals complexes like interaction energies, dipole moments, polarizabilities, etc. one should use the basis set augmented with diffuse function, because they allow to describe better the interactions of molecules that are far from each other (distances are larger than those of covalent bonds within molecules). Moreover, the use of a method accounting for the electron correlation (post HF) should be employed.

In case if we consider systems that have single reference character, at least the MP2 level of theory should be used, and the gold standard is the CCSD(T) level of theory. For multireference systems one could use the methods like CASPT2 or CASPT3 (but be sure there is a convergence of the perturbation theory), or MRCI.

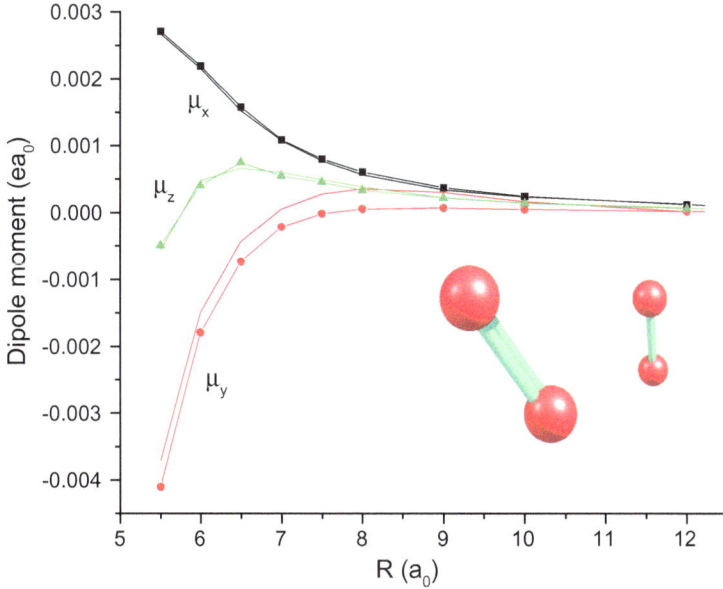

Fig. 3.1 Dipole moment components for the quintet state of O_2 dimer calculated at the UCCSD (T)/aug-cc-pVTZ level of theory with (*solid line + symbol*) and without (*solid line*) BSSE correction. The bond length r_{OO} was taken to be 2.29 a_0 for the first molecule and 2.38 a_0 for the second one

Electric properties of van der Waals complex when using ab initio calculations should be, if possible, corrected for the BSSE and BSIE similar to the interaction energy. The BSSE could reach up to 10 % for the dipole moments in the range of the van der Waals wells for the case of a medium sized basis set, like aug-cc-pVTZ (see Fig. 3.1, for example). However, for the systems with multireference character the BSSE correction is not used when a size-inconsistent method is employed.

3.2.1 Dipole Moments of Small Complexes

Dipole moments of atom-atomic complexes are now studied very well. These complexes are the simplest van der Waals complexes and, as a result, they have been studied first of all and fully enough (see, for instance [14–21]). Particular analytical forms of dipole moments for atom-atomic complexes can be easily obtained from the general expressions of Sect. 3.1 with a given accuracy using the symmetry properties (Appendix B) of interacting atoms. In particular, Eq. (3.1.9) gives for two interacted atoms (when at least one of them has nonzero quadruple moment) the asymptotic behavior as R^{-4}. And when these atoms are in the state s (spherical symmetry) the leading term ($\sim R^{-7}$) is caused only by the second dispersion term in (3.1.10). It should be noted that these dependences can be effectively used to construct the dipole moment function of diatomic molecules for large interatomic distances [22, 23].

The dipole moments of the atom-diatomic complexes X_2-Y, in contrast to the atom-atomic ones, depend also on the angle θ between the axis of diatomic molecule and the axis passing through the atom Y and the molecule X_2 (Fig. 3.2a). Note that if the nonrigidity of the diatomic molecule X_2 is taken into account, then the additional dependence of dipole moment of the complex on r appears. So, in general we have a surface of the dipole moment for such complex. Nevertheless, these van der Waals complexes are relatively simple yet and they are studied intensively up to now because of their importance (see, for instance, [24–30]).

X_2-Y complex. The r dependence can be easily obtained for X_2-Y complexes (where Y is an atom of noble gas in the ground state) in (3.1.9) if only the leading quadrupole-induced dipole interaction is taken into account. In this case, the equilibrium distances (R_e) of the complexes are comparable with the size of X_2 molecules. As a result, the modelling of a molecule in the form of a point, for which the interacting molecules are considered as not having the size, can not be applied and should be modified to take the size of a molecule into account. For this purpose, each molecule of a complex is considered as two effective point atoms without any interactions between them. The position of these atoms coincides with those of the nuclei of the molecule. The tensor of the total quadrupole moment of the effective atoms is the same as that of the molecule. For the diatomic homonuclear molecule the quadrupole moments of the effective atoms are equal to each other. Thus, the dipole moment of the complex is a function of intra- and intermolecular separations and relative orientation of the complex components. As a result, in the framework

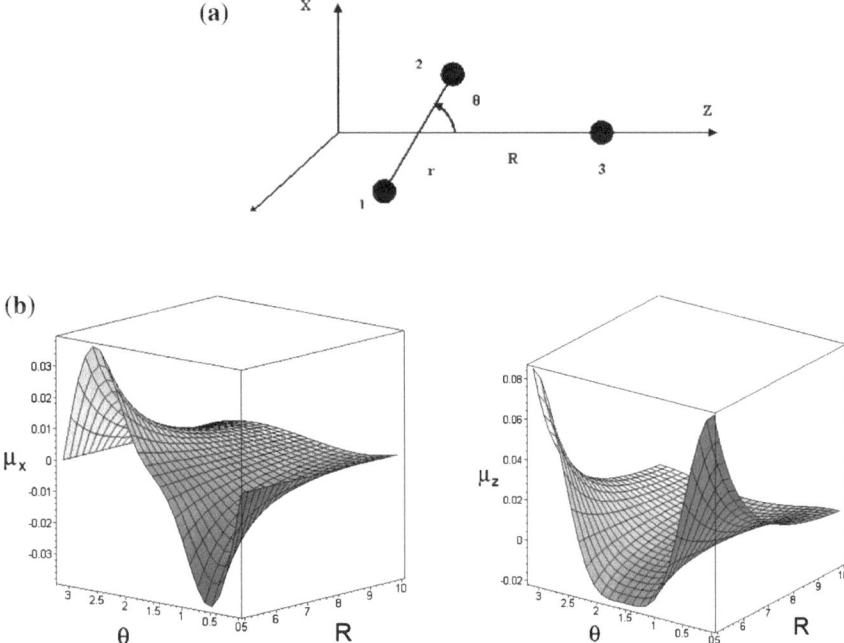

Fig. 3.2 **a** Atom-diatomic molecular complex X_2-Y. *1*, *2*—the atoms of X_2 molecule; *3*—the atom Y; *r* is the distance between atoms of X_2; *R* is the distance between Y and X_2. **b** Components of the dipole moment of N_2–Ar complex calculated in the framework of the modified induction model (R, μ_x and μ_y are in a.u., θ in rad)

of the modified model the components of the dipole moment of the complex may be written in the form

$$\mu_\alpha = \sum_{p<q} \mu_\alpha^{pq} \tag{3.2.1}$$

where the indexes p and q are the numbers of atoms in a complex and the values μ_α^{pq} are the dipole moment components of pq-bond. In this way the following expressions for dipole moment components can be obtained

$$\mu_x = \mu_y = \frac{\alpha\Theta_{zz}}{R^4} P_2^1(\cos\theta) + \frac{3}{2}\frac{\alpha\Theta_{zz}r^2}{R^6} P_4^1(\cos\theta) + \cdots, \tag{3.2.2}$$

$$\mu_z = -3\frac{\alpha\Theta_{zz}}{R^4} P_2^0(\cos\theta) - \frac{15}{2}\frac{\alpha\Theta_{zz}r^2}{R^6} P_4^0(\cos\theta) + \cdots. \tag{3.2.3}$$

where α is the polarizability of the inert atom Y, Θ_{jj}—quadrupole moment of X_2, $P_n^m(\cos\theta)$—associated Legendre polynomials. The first terms in Eqs. (3.2.2) and (3.2.3) correspond to the point model and the second terms give contribution into dipole moment of complex caused by size effect of the molecule X_2. In Fig. 3.2b the R and θ dependences of the dipole moment components for N_2-Ar complex are shown.

X_2-Y_2 complex. Analogous approach can also be applied to the X_2-X_2 and X_2-Y_2 complexes. In the modified model the tensors of polarizability and quadrupole moment of effective atoms X(Y) are equal each to other and their total polarizability and quadrupole moment are the same as that of the molecule. So, retaining the leading term, the dipole moment of the complex X_2-X_2 may be written as

$$\mu_\alpha = \sum_{p<q} (\mu_\alpha^{pq} - \mu_\alpha^{qp}) \tag{3.2.4}$$

where

$$\mu_l^{pq} = \frac{1}{R_{pq}^7} \sum_{ijk} \alpha_{il}^p \Theta_{jk}^q \left\{ 5R_{pq,i}R_{pq,j}R_{pq,k} - R_{pq}^2 R_{pq,i}\delta_{jk} - R_{pq}^2 R_{pq,k}\delta_{ij} - R_{pq}^2 R_{pq,j}\delta_{ik} \right\}. \tag{3.2.5}$$

In Eq. (3.2.5) α_{il}^p and Θ_{jk}^q are the polarizability and quadrupole moment tensors of pth and qth atoms respectively; R_{pq} is the separation between pth and qth atoms, $R_{pq,i}$ is the projection of the vector $\vec{R}_{pq}$ on the axis i and δ_{ij} is the Kronecker delta. The dipole moment surface in this case depends on R, r_1, r_2, θ_1, θ_2 and φ (see Fig. 3.3). Some calculations illustrate the dipole moment of the N_2–N_2 dimer in Table 3.1 and Fig. 3.4.

In order to demonstrate the performance of the long-range approximation (3.1.9) for the description of the dipole moment and the range of applicability of this model let us consider several complexes, such as CO_2–H_2, CO_2–CO_2 and N_2–H_2. The

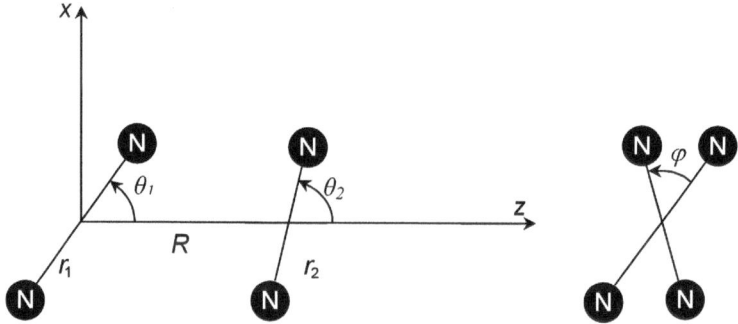

Fig. 3.3 Coordinate system of N_2-N_2 complex

Table 3.1 Dipole moment of N_2–N_2 dimer (in D)

Configuration	θ_1	θ_2	Φ	R_e, Å [115]	μ[115] $r_1 = r_2 = 2.08$ a_0	μ^a	μ^b	μ^c	μ^d	μ^e
Gear-45 (C_{2h})	45	45	0	4.09	0.000	0.000	0.000	0.000	0.000	0.000
Gear-30 (C_s)	60	30	0	4.10	0.017	0.019	0.023	0.023	0.023	0.022
Gear-15 (C_s)	75	15	0	4.13	0.030	0.032	0.039	0.038	0.038	0.037
T (C_{2V})	90	0	0	4.15	0.034	0.036	0.044	0.043	0.043	0.042
X30 (C_1)	90	30	90	4.03	0.025	0.028	0.034	0.034	0.034	0.033
X60 (C_1)	90	60	90	3.78	0.001	0.011	0.014	0.014	0.014	0.014
X (D_{2d})	90	90	90	3.63	0.000	0.000	0.000	0.000	0.000	0.000
Parallel (D_{2h})	90	90	0	3.70	0.000	0.000	0.000	0.000	0.000	0.000
Linear ($D_{\infty h}$)	0	0	0	5.22	0.000	0.000	0.000	0.000	0.000	0.000

Author's calculations for $r_1 = r_2 = 2.07$ a_0: [a]HF method with aug-cc-pVTZ basis set plus midbond functions (BSSE corrected); [b]MP2 method with aug-cc-pVTZ basis set plus midbond functions (BSSE corrected); [c]MP2 method with aug-cc-pVQZ basis set plus midbond functions (BSSE corrected); [d]CCSD(T) method with aug-cc-pVTZ basis set plus midbond functions (BSSE corrected); [e]Modified model

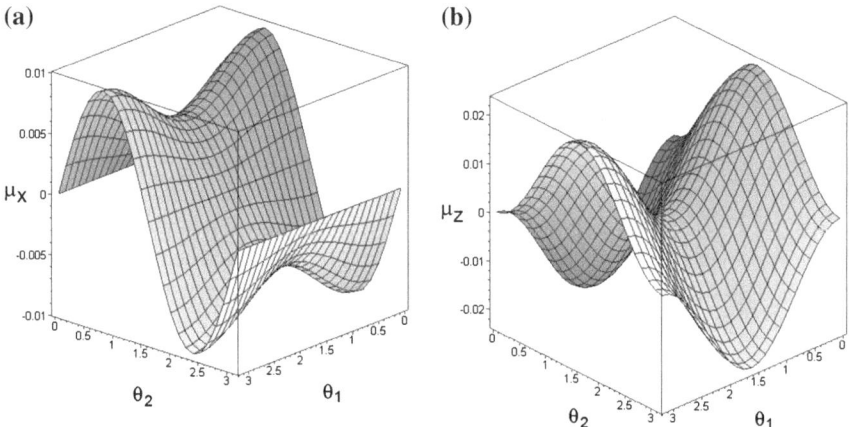

Fig. 3.4 Components of the dipole moment (ea_0) of N_2–N_2 complex calculated in the framework of the modified induction model (the angles are in rad). $R = 7.14$ a_0, $r_1 = r_2 = 2.07$ a_0, $\varphi = \pi/2$

bond lengths of monomers were kept frozen in their ground state geometry. We have limited the expansion of induction part (3.1.9) and dispersion part 3.1.17 of dipole moment to the terms $\sim R^{-4}$ through the terms $\sim R^{-7}$. The results of analytical representation and ab initio calculations for selected angular orientations are

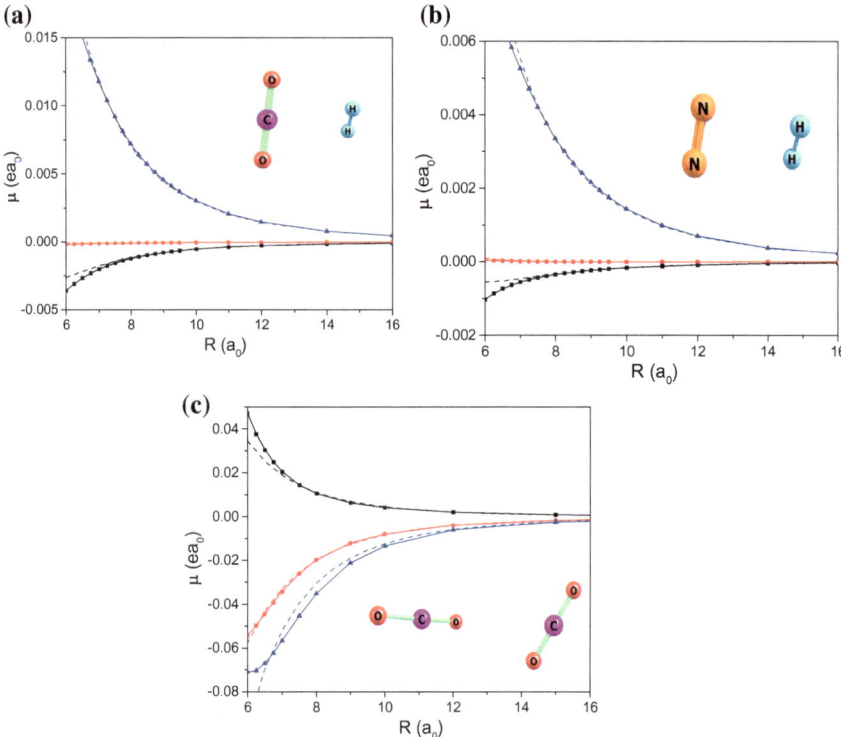

Fig. 3.5 Dipole moment of CO_2–H_2 (**a**), N_2–H_2 (**b**) and CO_2–CO_2 (**c**) complexes calculated in the framework long-range approximation (*dotted lines*) and at the CCSD(T) level of theory (*solid lines +symbols*). *Black color* μ_x; *red color* μ_y; *blue color* μ_z

presented at Fig. 3.5. The ab initio calculations were carried at for all cases using the finite-field method with the BSSE correction at the CCSD(T) level of theory in aug-cc-pVTZ basis set. The properties used for analytical calculations are reported in Tables 5.4 and 5.5. It is notable from the Fig. 3.5 that the long-range approximation for considered complexes reproduces well the dipole moment for $R > 7$ a_0. It should be noted from the figure that the z-component of the dipole moment is the largest because the z-axis is the axis along which the molecules interact here.

3.2.2   Dipole Moment Surface of the van der Waals Complex CH_4–Ar

In this paragraph we will concentrate on ab initio calculations of the dipole moment of methane-argon complex relevant to the atmospheres of Titan and Earth. Also, the analytical form of the dipole moment components is presented for the case of rigid methane molecule.

Table 3.2 Basis functions used for representation of dipole moment of the CH$_4$–Ar complex

μ_x, μ_y		μ_z	
l	m	L	m
1	1	1	0
2	1	2	2
3	1	3	0
3	3	3	2
4	1	4	2
4	3	4	4
5	1	5	0
5	3	5	2
5	5	5	4
6	1	6	2
6	3	6	4
6	5	6	6
7	1	7	0
7	3	7	2
7	5	7	4
7	7	7	6
8	1	8	2
8	3	8	4
8	5	8	6
8	7	8	8

The ab initio calculations were carried out at the CCSD(T) level of theory with aug-cc-pVTZ basis set augmented by midbond functions (for more details, see Ref. [31]). Then, the full 3D surface of each dipole moment component was represented as a sum of spherical harmonics multiplied by corresponding expansion coefficients and a cosine/sine function.

$$\mu_{x,z} = \sum_{l,m} (-1)^m a_{lm}(R) P_{lm}(\cos(\theta)) \cos(m\varphi),$$

$$\mu_y = \sum_{l,m} (-1)^m a_{lm}(R) P_{lm}(\cos(\theta)) \sin(m\varphi).$$

The possible values of l and m for the complex under consideration are reported in Table 3.2. The expansion coefficients $a_{lm}(R)$ were found by a least square fit at each point of R to ab initio results. Then, the coefficients were interpolated by cubic splines for the range of R: 4.5–30 a_0 covering both repulsive region and the region of van der Waals attraction. The relative errors of the fitted surfaces do not exceed 5 % for the range of short separations and 0.5 % for long-range separations.

3.2.3 Dipole Moment Surface of the van der Waals Complex CH_4-N_2

The description of dipole moments of molecular complexes is significantly sophisticated as the number of atoms in the interacting molecules increases. The electrical properties of molecular van der Waals complexes are now studied not fully enough. Below, we concentrate our attention only on two relatively large molecular complexes CH_4–N_2 and C_2H_2–C_2H_2 (Sect. 3.2.4) which have, on the one hand, the astrophysical interest and on the other hand, they illustrate effectively the theory discussed above.

It is well known that any gas media consisting of nonpolar molecules absorb in IR and far IR spectra [32, 33]. The nature of absorption is in the presence of both transient dipole moment of colliding molecules, and of the dipole moment of stable van der Waals complexes. For this reason, the dipole moment of interacting molecules is the object of numerous theoretical and experimental studies [24, 34–44]. One of the reasons arising an interest to the CH_4–N_2 complex is related to the study of the nitrogen-methane atmosphere of Titan [45–60] (the satellite of Saturn). Due to low temperatures (70–100 °C [51]) of its atmosphere and relatively high pressure (1.5 bar), the formation of stable van der Waals complexes CH_4–CH_4, N_2–N_2 and CH_4–N_2 is very probable. These complexes play an important role in all physical-chemical processes of Titan atmosphere [45, 46]. In particular, van der Waals complex CH_4–N_2 will manifest itself as spectral peculiarities on the background of non-resolved diffuse contours of absorption spectra of colliding molecules CH_4 and N_2 [33].

At present, the CH_4–N_2 complex is relatively well studied. There are several works [61–63], where the theoretical investigation of the potential energy surface was carried out. It was found that there is a family of the most stable configurations of the CH_4–N_2 complex with the energy difference of less then 0.04 cm^{-1} [63]. Electric properties of the complex were studied in Refs. [7, 64]. In Ref. [7] the analytical investigation of the long-range collision-induced dipole moment surface taking into account the induction (up to R^{-6}) and dispersion (up to R^{-7}) contributions was carried out. The dipole moment surface suggested in Ref. [7], was employed for the description of collisional spectra of molecules CH_4 and N_2 in works [65, 66]. In these works, it was shown that calculated absorption spectra for frequencies from 30 to 250 cm^{-1} agrees well with existing measurements [67, 68] but at the high frequencies >250 cm^{-1} it shows substantial intensity defect. In the work [64] (and Sect. 3.2.3) the long-range model of the dipole moment surface [64] was improved by fully including the induction terms up to R^{-7} and by accounting for the contributions from the effects of electron shells overlap of the interacting molecules.

A. Ab initio *and analytical calculations*

The calculation results of dipole moment surface of the van der Waals complex CH_4–N_2 are performed in this Section following to the theoretical treatment considered in previous Sect. 3.1. The symmetry properties of CH_4 and N_2 molecules (see Appendix B) allow us to simplify significantly general formulae obtained above.

In this book the Cartesian coordinate system shown in Fig. 3.6a is used for description of the complex. The origin of the coordinate system is placed on the carbon atom of the methane molecule. The vector R connects the carbon atom with the center of N_2 bond. It has the components $(R, 0, 0)$. The rotation of the methane (A) and the nitrogen (B) molecules in this coordinate system is determined by Euler angles $\chi_A, \theta_A, \varphi_A, \theta_B$ and φ_B, accordingly. The initial position of the molecules in the CH_4–N_2 complex presented in Fig. 3.6a corresponds to the Euler angles $\chi_A = \theta_A = \varphi_A = \theta_B = \varphi_B = 0$. In this figure the coordinates of nitrogen atoms are $(R, 0, \pm r_{NN}/2)$ and the CH_4 molecule has the standard orientation: the carbon atom is at the origin $(0, 0, 0)$ and hydrogen atoms have the coordinates (c, c, c), $(c, -c, -c)$, $(-c, -c, c)$ and $(-c, c, -c)$, where $c = r_{CH}/\sqrt{3}$. In the present study the geometries of the monomers CH_4 and N_2 were optimized at the CCSD(T)/aug-cc-pVTZ level of theory. The equilibrium bond lengths for the CH_4 and N_2 molecules were found to be $r_{CH} = 2.0596$ a_0 and $r_{NN} = 2.0864$ a_0, accordingly. During the calculations the monomers were kept rigid with geometric parameters mentioned above.

Ab initio calculations of the dipole moment components were carried out at the CCSD(T) level with the BSSE correction.

The mostly discussed in works [7, 61–64] configurations of the CH_4–N_2 complex, including the most stable configuration, are presented at Fig. 3.6b. The geometric parameters (Euler angles $\chi_A, \theta_A, \varphi_A, \theta_B$ and φ_B) for these configurations along with equilibrium distance R_e and interaction energy $\Delta E(R_e)$ are given in Table 3.2. The ab initio calculations of the dipole moment components μ_α for these configurations at the MP2 and CCSD(T) levels of theory are presented at Fig. 3.7.

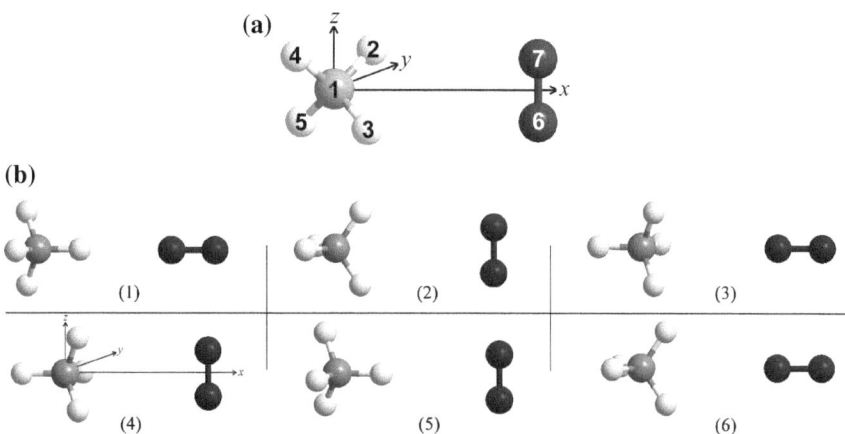

Fig. 3.6 a Coordinate system for the initial geometry; **b** General geometries of the CH_4–N_2 complex (Reprinted with permission from Ref. [64]. Copyright 2010 American Institute of Physics)

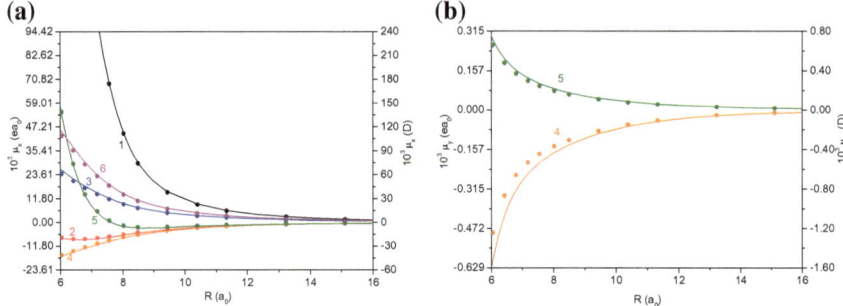

Fig. 3.7 Ab initio calculation of the dipole moment (**a**—μ_x, **b**—μ_y) for six configurations of the CH$_4$–N$_2$ complex (Fig. 3.6b). *Solid line* MP2 calculations; *circles* CCSD(T) calculations. The *numbers* indicate the configurations (see Table 3.3). (Reprinted with permission from Ref. [64]. Copyright 2010 American Institute of Physics)

Due to symmetry properties of the complex, the μ_x component exists for all configurations of the complex, μ_y component appears only for the configurations possesing C_S symmetry, and there is no μ_z component at all for the chosen six geometries. Figure 3.7a shows that with the R decreasing the μ_x component for the configurations 1, 3 and 6 (the N$_2$ molecule lies along the x-axis) monotonically goes up and for the configuration 4 monotonically goes down. The functions μ_x (R) for the configurations 2 and 5 have more complicated behaviour. The values of μ_y components are significantly smaller (by $\sim 10^2$) then the values of μ_x components because the molecules are arranged along the x-axis. The Fig. 3.7b shows that for the configuration 4 the function μ_y monotonically goes down and for the configuration 5 monotonically goes up when the molecules get closer. It should be noted that the MP2 and CCSD(T) calculations agree quite well .

For the interacting CH$_4$ (T_d symmetry) and N$_2$ ($D_{\infty h}$ symmetry) molecules, the induction dipole moment μ_{α}^{ind} through the order R^{-7} takes the form [5]

Table 3.3 Symmetry, geometrical parameters, equilibrium distance R_e and interaction energy $\Delta E(R_e)$ for general configurations (Fig. 3.6b) of the CH$_4$–N$_2$ complex

Configuration	Symmetry	χ_A	θ_A	φ_A	θ_B	φ_B	R_e (a$_0$) [63]	$\Delta E(R_e)$ (μE_h) [63]
1	C_{3V}	90	45	t	90	0	8.81	−412.251
2	C_{2V}	0	45	90	0	0	7.26	−566.445
3	C_{3V}	0	45	t	90	0	8.01	−369.226
4	C_S	0	45	t	0	0	6.84	−675.375
5	C_S	90	45	t	0	0	8.11	−291.741
6	C_{2V}	0	45	90	90	0	8.43	−319.591

The angles are in degrees
$t = (180/\pi) \arcsin(1/\sqrt{3})$

$$\mu_\alpha^{ind} = \frac{1}{3}\alpha_{\alpha\beta}^A\Theta_{\gamma\delta}^B T_{\beta\gamma\delta} - \frac{1}{9}A_{\alpha,\beta\gamma}^A\Theta_{\delta\varepsilon}^B T_{\beta\gamma\delta\varepsilon} + \frac{1}{15}\alpha_{\alpha\beta}^B\Omega_{\gamma\delta\varepsilon}^A T_{\beta\gamma\delta\varepsilon}$$

$$+ \frac{1}{105}\alpha_{\alpha\beta}^A\Phi_{\gamma\delta\varepsilon\varphi}^B T_{\beta\gamma\delta\varepsilon\varphi} - \frac{1}{105}\alpha_{\alpha\beta}^B\Phi_{\gamma\delta\varepsilon\varphi}^A T_{\beta\gamma\delta\varepsilon\varphi} + \frac{1}{45}E_{\alpha,\beta\gamma\delta}^A\Theta_{\varepsilon\varphi}^B T_{\beta\gamma\delta\varepsilon\varphi}$$

$$- \frac{1}{315}A_{\alpha,\beta\gamma}^A\Phi_{\delta\varepsilon\varphi\nu}^B T_{\beta\gamma\delta\varepsilon\varphi\nu} + \frac{1}{225}E_{\alpha,\beta\gamma\delta}^B\Omega_{\varepsilon\varphi\nu}^A T_{\beta\gamma\delta\varepsilon\varphi\nu} + \frac{1}{3}\alpha_{\alpha\beta}^B\Theta_{\gamma\delta}^B\alpha_{\varepsilon\varphi}^A T_{\beta\varepsilon}T_{\varphi\gamma\delta}$$

$$- \frac{1}{315}D_{\alpha,\beta\gamma\delta\varepsilon}^A\Theta_{\varphi\nu}^B T_{\beta\gamma\delta\varepsilon\varphi\nu}.$$

$$(3.2.6)$$

The dispersion contribution to the dipole moment through the order R^{-7} is defined by the following expression within the CRA2 approximation:

$$\mu_\alpha^{disp} = \frac{5\beta_{\alpha\beta\gamma}^A\alpha_{\delta\varepsilon}^B}{36\alpha^A\alpha^B}T_{\beta\delta}T_{\gamma\varepsilon}C_6 - \left(B_{\alpha\beta,\gamma\delta}^A\alpha_{\varepsilon\varphi}^B - B_{\alpha\beta,\gamma\delta}^B\alpha_{\varepsilon\varphi}^A\right)\frac{5T_{\beta\varepsilon}T_{\varphi\gamma\delta}C_6}{54\alpha^A\alpha^B}. \qquad (3.2.7)$$

The analytical expression for exchange dipole moment μ_α^{exch} is defined here by Eq. (3.1.21), where [13]

$$\delta = \frac{1}{\beta_A} + \frac{1}{\beta_B} + \frac{1}{2(\beta_A + \beta_B)} + 1, \qquad (3.2.8)$$

$$\eta = \frac{3}{4}\beta_A + \frac{3}{4}\beta_B. \qquad (3.2.9)$$

Here and thereafter, the parameters β_A and β_B are determined using the ionization potentials $U_A = \beta_A^2/2$ and $U_B = \beta_B^2/2$ of the interacting molecules A and B respectively. The applicability of analytical description of the dipole moment of the CH_4–N_2 complex in the framework of suggested model is illustrated in Fig. 3.8. In this figure the long-range analytical calculations, analytical calculations with taking into account the exchange contribution and the CCSD(T) calculations of the dipole moment components for configurations 3, 4 and 5 are given. The dispersion coefficient $C_6 = 96.94 \ E_h a_0^{-6}$ for interacting methane and dinitrogen molecules is taken from Ref. [69]. The parameters used for analytical calculations are given in [70–73] (see also Table 5.2). The exchange contribution to the dipole moment for considered configurations was found by fitting μ_α^{exch} to the difference between ab initio and long-range calculations in the range of potential well for each configuration. The obtained parameters B_α for six configurations (Fig. 3.6b) are presented in Table 3.4. The analysis of Fig. 3.8 shows that this approach describes well the dipole moment for the whole range of potential well of considered configurations, while the long-range approximation provides good results for $R > 10$ borhs. A noticeable divergence of analytical and ab initio dipole moment appears in the range of R outside of the well ($R < 7.4 \ a_0$ for configuration 3, $R < 6.3 \ a_0$ for configuration 4 and $R < 7.5 \ a_0$ for configuration 5 [63]). The Fig. 3.8 shows also

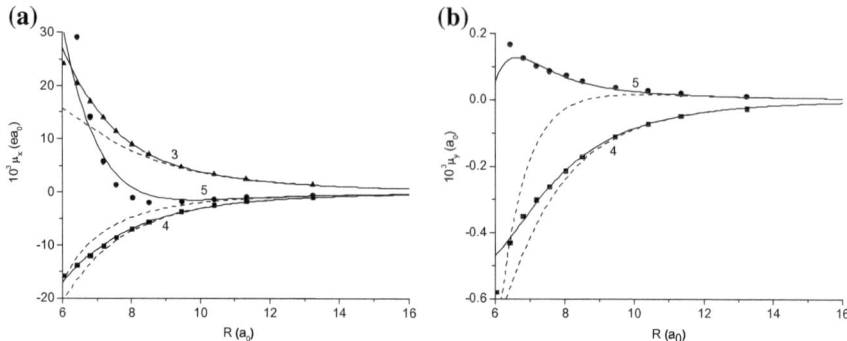

Fig. 3.8 Dipole moment (**a** μ_x, **b** μ_y) as a function of R for configurations 3, 4 and 5 of the complex CH$_4$–N$_2$ [13]. The *numbers* indicate the configurations. *Solid lines* analytical calculations with the exchange contribution; *dash lines* analytical calculations without the exchange contribution; CCSD(T) calculations: *triangles* configuration 3; *squares* configuration 4; *rounds* configuration 5. (Reprinted with permission from Ref. [64]. Copyright 2010 American Institute of Physics)

Table 3.4 Coefficients B_α (in a.u.) for the CH$_4$–N$_2$ dipole moment[a] [13]

Conf.	B_x	B_y
1	5.4669	–
2	0.4041	–
3	0.3720	–
4	0.1240	0.00746
5	1.5859	0.02739
6	0.2521	–

[a]$\mu_\alpha^{exch} = B_\alpha R^{3.20238} e^{-1.53525R}$

that analytical dipole moment without the exchange contribution leads to incorrect (in some cases to dramatic) behaviour of dipole moment in the range $R < R_e$. One should note, that there is also a very good agreement between ab initio and analytical calculations (with μ_α^{exch}) for configurations 1, 2 and 6 in the range of potential well.

The complete analytical description of the dipole moment in the framework of considered model requires the knowledge of the angular dependence of $B_\alpha(\chi_A, \theta_A, \varphi_A, \theta_B, \varphi_B)$ in Eq. (3.1.21). The possibility of the analytical description of μ_α can be illustrated using the particular examples (rotation of CH$_4$ by angle φ_A, and rotation of N$_2$ by angle θ_B). The numerical values of coefficients B_α for 42 complex configurations appeared in this way (13 configurations obtained by rotation of angle θ_B by 15° from 0° to 180° with fixed coordinates of CH$_4$; 29 configuration obtained by rotation of angle φ_A from 0° to 360° with fixed coordinates of N$_2$) were obtained as described above and are presented in Fig. 3.9.

Figure 3.9 illustrates the dependence of coefficients B_α on the angle θ_B for fixed Euler angles $\chi_A = 0°$, $\theta_A = 45°$, $\varphi_A = (180/\pi) \arcsin(1/\sqrt{3})$ and $\varphi_B = 0°$. It is

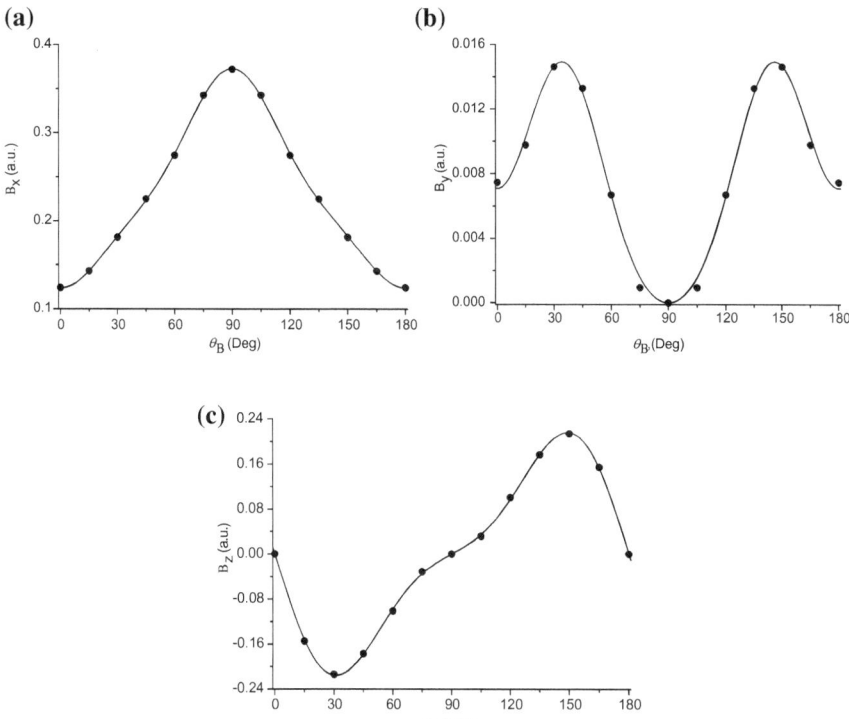

Fig. 3.9 Dependence of the coefficients B_α on the angle θ_B for the configurations of the CH$_4$–N$_2$ complex with fixed $\chi_A = 0°$, $\theta_A = 45°$, $\varphi_A = (180/\pi)$ arcsin $(1/\sqrt{3})$ and $\varphi_B = 0°$ [13]. Points— values of B_α deduced from the ab initio calculation (see text); lines—analytical calculation by Eqs. (3.2.10) and (3.2.11) (Reprinted with permission from Ref. [64]. Copyright 2010 American Institute of Physics.)

noticable, that the numerical values of B_α form the smooth functions depending on the angle θ_B, which can be expanded in series in terms of Legendre polynomials of low order. The analytical expressions for the coefficients B_α, obtained by fitting to the numerical values deduced from the ab initio calculation, have the forms (in a.u.):

$$B_x(\theta_B) = 0.2732 - 0.1601 P_2^0(\cos\theta_B) + 0.0333 P_4^0(\cos\theta_B) - 0.0227 P_6^0(\cos\theta_B),$$
(3.2.10)

$$B_y(\theta_B) = 0.00696 + 0.01060 P_2^0(\cos\theta_B) - 0.00718 P_4^0(\cos\theta_B) - 0.00330 P_6^0(\cos\theta_B),$$
(3.2.11)

$$B_z(\theta_B) = -0.1076 P_2^1(\cos\theta_B) - 0.0307 P_4^1(\cos\theta_B) - 0.0011 P_6^1(\cos\theta_B).$$ (3.2.12)

Here $P_l^m(\cos\theta_B)$ are the associated Legendre polynomials. It should be pointed out, that at $\theta_B = 0°$ the complex CH$_4$–N$_2$ is in configuration 4 and at $\theta_B = 90°$ in

configuration 3, and the B_α values in Eqs. (3.2.10) and (3.2.11) for these configurations $(B_x(0°) = 0.1237$ a.u., $B_x(90°) = 0.3728$ a.u. and $B_y(0°) = 0.00708$ a.u.)
are in a good agreement with the values in Table 3.4. It is obvious, that the
functions B_α in Eqs. (3.2.10) and (3.2.11) are periodic ones with the period of 180°
due to the homonuclearity of the N_2 molecule.

The dependence of the coefficients B_α on the angle φ'_A $(\varphi'_A = \varphi_A - 90°)$ for
fixed $\chi_A = 0°$, $\theta_A = 45°$, $\theta_B = 0°$ and $\varphi_B = 0°$ is shown at Fig. 3.10.

The angle φ'_A is introduced to show the symmetry properties of the complex (and
the coefficients B_α, respectively) more cleary when the the molecule CH_4 is rotated
by the angle φ_A. For the considered configurations of the CH_4–N_2 complex the
coefficient B_α has only x and y components and has the properties: $B_x(180° -
\varphi'_A) = B_x(180° + \varphi'_A)$, $B_y(180° - \varphi'_A) = -B_y(180° + \varphi'_A)$. The analytical expres-
sions for these coefficients have the following form (in a.u.)

$$B_x(\varphi'_A) = 0.6508 - 0.6558P_1^0(\cos \varphi'_A) - 0.0515P_2^0(\cos \varphi'_A) + 0.6386P_3^0(\cos \varphi'_A)$$
$$- 0.1913P_4^0(\cos \varphi'_A) - \quad -0.0655P_5^0(\cos \varphi'_A) - 0.0778P_6^0(\cos \varphi'_A)$$

$$(3.2.13)$$

$$B_y(\varphi'_A) = -0.01062P_1^1(\cos \varphi'_A) - 0.03098P_2^1(\cos \varphi'_A) + 0.01854P_3^1(\cos \varphi'_A)$$
$$- 0.02633P_4^1(\cos \varphi'_A) \quad - 0.00127P_5^1(\cos \varphi'_A) - 0.00485P_6^1(\cos \varphi'_A).$$
$$+ 0.00185P_7^1(\cos \varphi'_A)$$

$$(3.2.14)$$

The coefficients B_α in Eqs. (3.2.13) and (3.2.14) are the periodic functions of φ_A
with the period of 360°. The coefficients $B_\alpha(\varphi'_A)$ (and $B_\alpha(\varphi_A)$, correspondently) are
also the smooth functions. So, the considered particular cases of molecular rotations

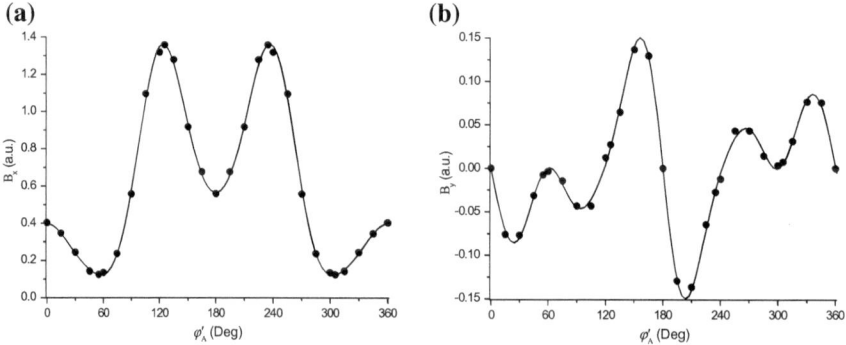

Fig. 3.10 Dependence of the coefficients B_α on the angle φ'_A for fixed $\chi_A = 0°$, $\theta_A = 45°$, $\theta_B = 0°$
and $\varphi_B = 0°$ for the methane—dinitrogen complex [13]. (Reprinted with permission from Ref.
[64]. Copyright 2010 American Institute of Physics)

in the complex show that the exchange contributions into the dipole moment of the complex CH_4–N_2 may be obtained in the analytical form.

B. *Dipole moment of the most stable configuration of CH₄–N₂ complex*

It is also interesting to consider the dipole moment of the most stable configuration of CH_4–N_2 complex. In Ref. [63] it was found that the CH_4–N_2 complex has a family of such configurations with practically equal interaction energies (the difference is less than 0.04 cm^{-1}). These configurations can be obtained from the configuration 4 by rotation of the N_2 molecule by angle τ over the x axis at fixed equilibrium intermolecular separation $R_e = 6.8$ a_0. The rotation of nitrogen molecule by angle τ corresponds to its rotation by angle θ_B at fixed angle $\varphi_B = 90°$.

The dipole moment of the complex was obtained for angle τ from $0°$ to $180°$ by $15°$ (13 configurations). For the calculations we have employed Eq. (3.1.2), that provides enough accuracy for the correct description of small changing in $\mu_x(R, \tau)$. The resulting analytical form for $\mu_x(R, \tau)$ was found to be as follows (in a.u.)

$$\mu_x(R, \tau) = -\frac{27.67779}{R^4} - \frac{78.76850}{R^5} + \frac{379.8549}{R^6} + \frac{687.0685}{R^7} + B_x(\tau)f(R),$$

$$(3.2.15)$$

$$\mu_y(R, \tau) = -\left(\frac{8.282}{R^5} + \frac{39.795}{R^6}\right)\left(2\cos^2\tau - 1\right)$$
$$+ \frac{57.053\cos^4\tau + 610.461\cos^2\tau - 326.625}{R^7} + B_y(\tau)f(R),$$

$$(3.2.16)$$

$$\mu_z(R, \tau) = -\left(\frac{16.564}{R^5} + \frac{79.590}{R^6} + \frac{57.053\cos^2\tau - 696.041}{R^7}\right)\sin\tau\cos\tau + B_z(\tau)f(R),$$

$$(3.2.17)$$

where

$$f(R) = R^{3.20238}\exp(-1.53525R)\cdot \qquad (3.2.18)$$

The fitting parameters $B_\alpha(\tau)$ were obtained from ab initio values of dipole moments at $R_e = 6.8$ a$_0$ and can be written as

$$B_x(\tau) = 0.112522 + 0.000154\cos^2\tau\left(4\cos^2\tau - 3\right)^2, \qquad (3.2.19)$$

$$B_y(\tau) = -0.005913 + 0.000111\cos^2\tau + 0.022825\cos^4\tau - 0.008514\cos^6\tau,$$

$$(3.2.20)$$

$$B_z(\tau) = \left(0.019364 - 0.007734\cos^2\tau\right)\sin\tau\cos\tau\cdot \qquad (3.2.21)$$

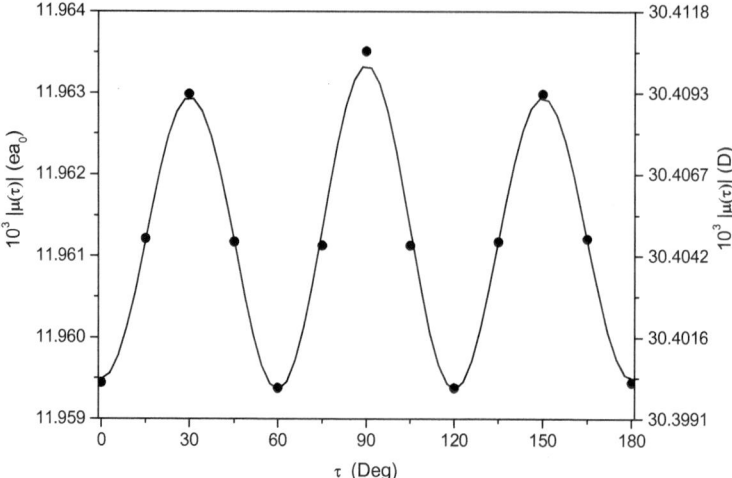

Fig. 3.11 Angular dependence (τ) of the dipole moment modulus for the most stable configurations of the CH$_4$–N$_2$ complex. (Reprinted with permission from Ref. [64]. Copyright 2010 American Institute of Physics.)

The results of ab initio and analytical calculations of dipole moment at R_e are presented in Fig. 3.11. It should be noted, that the major contribution (~ 96 %) to μ is the induction one, the dispersion (~ 16 %) and exchange (~ 12 %) contributions have opposite signs and partially cancel out each other. The dipole moment components $\mu_y(\tau)$ and $\mu_z(\tau)$ are significantly smaller then $\mu_x(\tau)$. The Fig. 3.11 shows, that the modulus of the dipole moment $|\mu(\tau)| = \sum_\alpha \mu_\alpha^2(\tau)$ of the CH$_4$–N$_2$ complex being in the most stable configurations is weakly dependent on angle τ ($|\mu(\tau)| = 0.011961 \pm \Delta|\mu(\tau)|$ ea_0 where the variations $\Delta|\mu(\tau)| < 0.000003$ ea_0). The behaviour of the modulus $|\mu(\tau)|$ is similar to the behaviour of $\mu_x(\tau)$ component. It is interesting to note, that a very weak dependence on angle τ is also observed for the polarizability invariants of the complex CH$_4$–N$_2$ (see next chapter and Ref. [64]).

In this Section, the dipole moment of the CH$_4$–N$_2$ complex was obtained using both ab initio and analytical methods. The analysis of the ab initio and analytical results has shown that the long-range model describes well the dipole moment for $R > 10$ a_0. For smaller R, when electron shells of interacting molecules begin to overlap, the dipole moment of complex can't be described correctly using the long-range approximation (even including the higher order terms of perturbation theory). However, for small overlap of electron shells when the exchange interactions are still small and the long-range approximation is weakly broken (that's the range of potential well of the van der Waals complex) there is a chance to describe the dipole moment of the complex in the analytical form. For even smaller R, when overlapping of the valence electrons of interacting molecules becomes significant, the numeric quantum mechanical calculations are only possible.

It should be noted that the suggested model allowing to take into account the effects related to the overlap of shells of valence electrons is based on the point model of interacting molecules and can be applied only to small molecules. Such limitation is due to the fact, that the parameters of exchange interaction β_A and β_B are defined only by the ionization potentials of molecules and take into account neither the form nor the size of interacting molecules.

3.2.4 Dipole Moment Surface of the Ethylene Dimer

Ethylene dimer in contrast to the CH_4–N_2 complex is studied significantly wider. The history of quantum-mechanical calculations of the interaction energy of two ethylene molecules begins with the work of Hashimoto and Isobe in 1973 [74]. Since then a number of theoretical works [75–94] have been devoted to the ab initio calculations of the potential energy of the dimer. The other approach to the investigation of the potential energy surface of the C_2H_4–C_2H_4 complex was applied in Refs. [95–99] in the framework of analytical description of long-range interactions between two ethylene molecules. This approach gives a physically correct analytical description of the potential energy surface for interacting C_2H_4 molecules at large intermolecular separations, R. However, in the framework of this approach, the well depths of different configurations of the C_2H_4–C_2H_4 dimer can not be described correctly, because at these intermolecular separations the electron shells of interacting ethylene molecules begin to overlap and the exchange interactions start to play an important role.

In contrast to the potential energy, the dipole moment of the ethylene dimer has been investigated not sufficiently. It is known that the dipole moment of ethylene dimer being in the most stable configuration (possessing symmetry D_{2d}) equals to zero. The theoretical calculations of the dipole moment surface of ethylene dimer have been carried out in the work [100]. There are also experimental works devoted to the collision-induced absorption in ethylene in the infrared region [101–104] that contain the information on the dipole moment surface of the interacting ethylene molecules. These collision-induced absorption spectra were used only for the evaluation of the quadrupole moment of single ethylene molecule.

In this section, following to the work [100] the results of the analytical and high-level ab initio calculations of the dipole moment for selected configurations of the ethylene dimer are discussed. For the analytical calculations of the dipole moment the monomer properties calculated in Ref. [100] are used (see Table 3.5).

A. Results of the PES and DMS calculations

For the accurate ab initio study of potential energy and dipole moment of the C_2H_4–C_2H_4 complex the methods considering the electron correlation should be used. For this reason, the following quantum-mechanical methods were employed: CCSD(T), CCSD(T)-F12, and MP2. All ab initio calculations in this Section were

Table 3.5 Molecular properties for C_2H_4 molecule

Property	[100][a]	Lit.	Property	[100][a]	Lit.
Θ_{xx}	−2.47	−2.42[b]	$C_{xx,xx}$	72.19	51.93[d]
Θ_{yy}	1.25	1.23[b]	$C_{xx,yy}$	−19.88	−10.65[d]
Θ_{zz}	1.22	1.19[b]	$C_{xx,zz}$	−52.31	−41.27[d]
Φ_{xxxx}	18.90	18.42[b]	$C_{xy,xy}$	45.35	29.24[d]
Φ_{yyyy}	−17.02	−16.27[b]	$C_{xz,xz}$	75.40	58.71[d]
Φ_{zzzz}	−15.89	−15.51[b]	$C_{yy,yy}$	75.82	51.47[d]
α_{xx}	22.05	22.41[c]	$C_{yy,zz}$	−55.95	−40.82[d]
α_{yy}	24.96	25.21[c]	$C_{yz,yz}$	118.68	101.00[d]
α_{zz}	34.24	34.24[c]	$C_{zz,zz}$	108.26	82.09[d]
$E_{x,xxx}$	−65.27	−80.67[d]	$B_{xx,xx}$	−733.70	–
$E_{y,yyy}$	−92.39	−81.01[d]	$B_{xx,yy}$	314.70	–
$E_{z,zzz}$	71.29	88.90[d]	$B_{xx,zz}$	419.00	–
$E_{x,xyy}$	−8.81	−8.11[d]	$B_{xy,xy}$	−399.39	–
$E_{x,xzz}$	74.08	88.78[d]	$B_{xz,xz}$	−574.40	–
$E_{z,zxx}$	−78.43	−62.24[d]	$B_{yy,yy}$	−394.06	–
$E_{z,zyy}$	7.14	−26.63[d]	$B_{yy,zz}$	188.14	–
$E_{y,yxx}$	−75.09	−88.80[d]	$B_{yz,yz}$	−591.64	–
$E_{y,yzz}$	167.48	169.82[d]	$B_{zz,zz}$	−579.30	–

[a]Calculated at the CCSD(T)/aug-cc-pVTZ level of theory using approach described in Ref. [105]
[b]Reference [106]; [c]Reference [107]; [d]Reference [96]

carried out with the correlation-consistent aug-cc-pVTZ basis set. For the evaluation of binding energy in the CBS limit, the aug-cc-pVQZ basis set was also used. In order to correct the energy for the basis set incompleteness error (BSIE), the energy was extrapolated to the CBS limit using the schemes of Martin (2.3.11) and Helgaker (2.3.12). The dipole moment of the ethylene dimer was calculated by the finite-field method described by Cohen and Roothaan [1] using Eq. (3.1.1) which provides enough accuracy for the system under consideration (the errors do not exceed 2×10^{-5} ea_0). The experimental ground state geometry (averaged over zero-point vibrations) of ethylene molecule was used for all calculations: the bond lengths and angles of ethylene molecule $r_{CC} = 2.5303$ a_0, $r_{CH} = 2.0504$ a_0, $\angle HCC = 121.085°$ and $\angle HCH = 117.83°$ were taken from Ref. [108]. The coordinate system of C_2H_4–C_2H_4 complex is shown in Fig. 3.12.

The long-range analytical approximation used for the description of dipole moment of the ethylene dimer, following the Sect. 3.1.2, gives the induction and dispersion contributions through the order R^{-7} as (here, the exchange interactions are not considered)

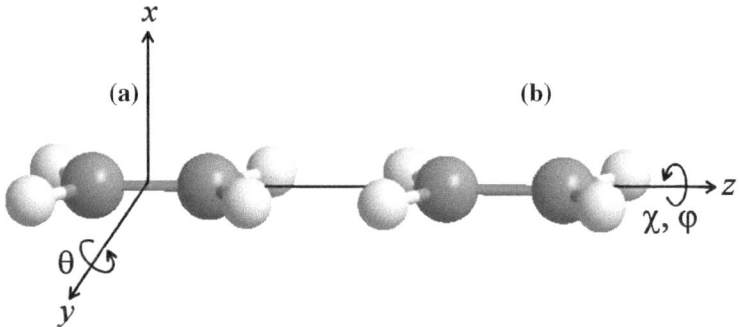

Fig. 3.12 Coordinate system of the C_2H_4–C_2H_4 complex (Reprinted with permission from Ref. [100]. Copyright 2011 Wiley Periodicals, Inc.)

$$\mu_{\alpha,ind} = \frac{1}{3} T_{\beta\gamma\delta} \left(\alpha^A_{\alpha\beta} \Theta^B_{\gamma\delta} - \alpha^B_{\alpha\beta} \Theta^A_{\gamma\delta} \right) + \frac{1}{105} T_{\beta\gamma\delta\varepsilon\varphi} \left(\alpha^A_{\alpha\beta} \Phi^B_{\gamma\delta\varepsilon\varphi} - \alpha^B_{\alpha\beta} \Phi^A_{\gamma\delta\varepsilon\varphi} \right)$$
$$+ \frac{1}{45} T_{\beta\gamma\delta\varepsilon\varphi} \left(E^A_{\alpha,\beta\gamma\delta} \Theta^B_{\varepsilon\varphi} - E^B_{\alpha,\beta\gamma\delta} \Theta^A_{\varepsilon\varphi} \right) - \frac{1}{3} T_{\beta\varepsilon} T_{\varphi\gamma\delta} \left(\alpha^A_{\alpha\beta} \Theta^A_{\gamma\delta} \alpha^B_{\varepsilon\varphi} - \alpha^B_{\alpha\beta} \Theta^B_{\gamma\delta} \alpha^A_{\varepsilon\varphi} \right)$$

$$(3.2.22)$$

and

$$\mu_{\alpha,disp} = \frac{5C_6}{54\alpha^2} T_{\beta\varepsilon} T_{\varphi\gamma\delta} \left(B^A_{\alpha\beta,\gamma\delta} \alpha^B_{\varepsilon\varphi} - B^B_{\alpha\beta,\gamma\delta} \alpha^A_{\varepsilon\varphi} \right).$$

$$(3.2.23)$$

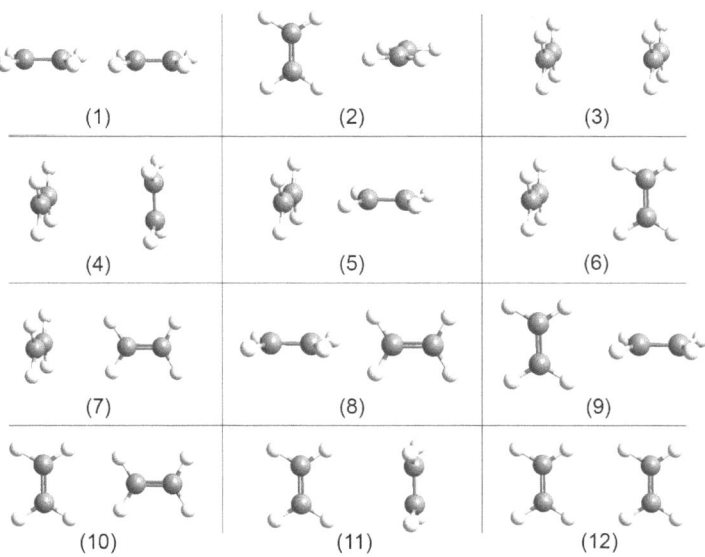

Fig. 3.13 Selected configurations of the C_2H_4–C_2H_4 complex (Reprinted with permission from Ref. [100]. Copyright 2011 Wiley Periodicals, Inc.)

Table 3.6 Equilibrium distance R_e (in a_0) and binding energy $\Delta E(R_e)$ for selected configurations of C_2H_4–C_2H_4 complex calculated at the CCSD(T) and CCSD(T)-F12 levels of theory with the BSSE correction in aug-cc-pVTZ basis set [100]

Conf.	Sym.	χ_A	θ_A	φ_A	χ_B	θ_B	φ_B	R_e	$\Delta E(R_e)$ CCSD(T)/ aug-cc-pVTZ	$\Delta E(R_e)$ CCSD(T)-F12/ aug-cc-pVTZ	$\Delta E_{CBS}^{CCSD(T)}(R_e)$ Helgaker	$\Delta E_{CBS}^{CCSD(T)}(R_e)$ Martin
1	D_{2h}	0	0	0	0	0	0	9.83	−132.23	−140.07	−141.82	−141.10
2	D_{2d}	90	90	90	90	0	90	7.18	−480.91	−509.96	−517.24	−514.20
3	D_{2h}	0	0	90	90	90	90	8.50	−22.15	−28.40	−29.84	−30.19
4	D_{2d}	0	0	90	90	90	0	8.50	−19.76	−25.59	−27.69	−28.02
5	C_{2v}	0	0	90	0	90	0	8.69	−210.66	−221.92	−225.76	−224.31
6	C_{2v}	0	90	90	90	90	0	7.18	−345.19	−366.67	−374.85	−372.45
7	C_{2v}	0	0	90	0	90	0	8.69	−201.70	−212.49	−216.52	−215.03
8	D_{2d}	0	0	0	0	0	90	9.45	−219.75	−233.52	−236.54	−235.20
9	C_{2v}	90	0	90	0	0	0	8.31	−336.51	−356.19	−360.98	−359.04
10	C_{2v}	90	0	90	0	0	90	9.07	−147.94	−155.03	−156.82	−156.25
11	C_{2v}	90	0	90	90	0	0	7.37	−361.99	−380.75	−388.02	−386.04
12	D_{2h}	90	90	90	90	0	0	8.31	−140.43	−146.65	−149.49	−148.93

The binding energies in the CBS limit calculated at the CCSD(T) level $\left(\Delta E_{CBS}^{CCSD(T)}(R_e)\right)$ using extrapolation schemes of Helgaker and Martin, are presented in the last two columns. All angles are in degrees, the energies are in cm^{-1}

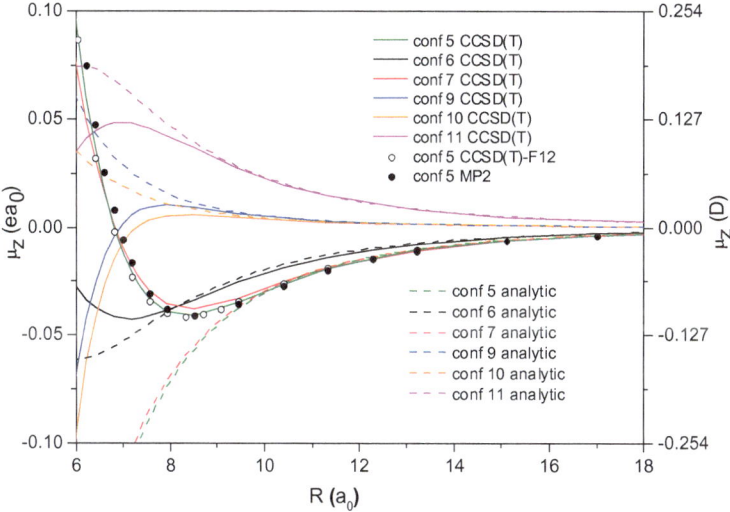

Fig. 3.14 Interaction-induced dipole moment component μ_z of the C_2H_4–C_2H_4 complex [100]. *Solid lines* calculations at the CCSD(T)/aug-cc-pVTZ level of theory with the BSSE correction; *dash lines* analytical calculations; *black circles* calculation at the MP2/aug-cc-pVTZ level of theory with the BSSE correction (Reprinted with permission from Ref. [100]. Copyright 2011 Wiley Periodicals, Inc.)

The calculation results are shown in Fig. 3.14 for 6 configurations of the complex with C_{2v} symmetry (configurations 5, 6, 7, 9, 10, and 11, see Fig. 3.13 and Table 3.6) with nonzero dipole moment (there is only μ_z component). It should be noted, that the MP2 level of theory gives slightly worse results compared to the "gold standard" CCSD(T) ones. Figure 3.14 shows also that the analytical description of the dipole moment provides a good agreement with the ab initio calculations at the CCSD(T)/aug-cc-pVTZ level of theory for $R > 9\ a_0$.

The analytical calculations of the dipole moment surface were carried out using the multipole moments and polarizabilities given in Table 3.5 (Ref. [100]) and the coefficient $C_6 = 300.2\ E_h a_0^{-6}$ taken from Ref. [109]. It should be noted, that the calculated multipole moments and polarizabilities of single ethylene molecule (Table 3.5) are in a good agreement with those from Refs. [96, 107, 109]. The analytical approach allows to evaluate different long-range contributions to the total dipole moment of the ethylene dimer. For example, the dipole moment contributions for configurations 5 and 9 of the complex with large and small interaction-induced dipole moments have been calculated. The results of calculations are shown in Fig. 3.15. As expected [see Eqs. (3.2.6) and (3.2.7)], the induction contribution to the dipole moment of the complex is significantly larger than the dispersion contribution. Moreover, the dipole polarizability-quadrupole induction terms $\left(\alpha^A \Theta^B + \alpha^B \Theta^A\right)$ give the major contribution to the dipole moment for all configurations where the long-range approximation is fulfilled.

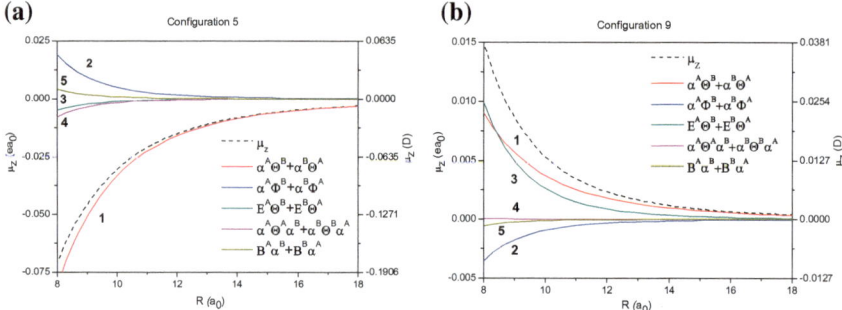

Fig. 3.15 Contributions to the dipole moment of configurations 5 (**a**) and 9 (**b**) of the C_2H_4–C_2H_4 complex. *Solid lines 1* induction contribution of order $R^{-4}\left(\alpha^A\Theta^B + \alpha^B\Theta^A\right)$, *2* induction contribution of order $R^{-6}\left(\alpha^A\Phi^B + \alpha^B\Phi^A\right)$, *3* induction contribution of order $R^{-6}\left(E^A\Theta^B + E^B\Theta^A\right)$, *4* induction contribution of order $R^{-7}\left(\alpha^A\Theta^A\alpha^B + \alpha^B\Theta^B\alpha^A\right)$, *5* dispersion contribution of order $R^{-7}(B^A\alpha^B + B^B\alpha^A)$; *dash lines* total dipole moment in the framework of long-range approximation. (Reprinted with permission from Ref. [100]. Copyright 2011 Wiley Periodicals, Inc.)

Analytical description of the DMS of the ethylene dimer allows us to understand some features this surface. In Fig. 3.14 we can see a similar behaviour of the dipole moment for configurations 5 and 7 and for configurations 9 and 10. It should be noted that the absolute values of the dipole moment functions for configurations 6 and 11 are also very close (there is only deference in the signs). Such behaviour of the dipole moment functions can be explained by the fact that the induction contributions $\alpha^A\Theta^B + \alpha^B\Theta^A$, $\alpha^A\Phi^B + \alpha^B\Phi^A$ and $\alpha^A\Theta^A\alpha^B + \alpha^B\Theta^B\alpha^A$ have the same values for configurations 5 and 7, for configurations 9 and 10 and differ only in sign for configurations 6 and 11. The small differences observed for these pairs of the dipole moment functions for large R are due to the difference in the induction $\left(E^A\Theta^B + E^B\Theta^A\right)$ and dispersion $(B^A\alpha^B + B^B\alpha^A)$ contributions.

References

1. H.D. Cohen, C.C.J. Roothaan, Electric dipole polarizability of atoms by the Hartree-Fock method. I. Theory of closed-shell systems. J. Chem. Phys. **43**(10), S34–S39 (1965)
2. M.J. Frisch, G.W. Trucks, H.B. Schlegel, G.E. Scuseria, M.A. Robb, J.R. Cheeseman, G. Scalmani, V. Barone, B. Mennucci, G.A. Petersson, H. Nakatsuji, M. Caricato, X. Li, H. P. Hratchian, A.F. Izmaylov, J. Bloino, G. Zheng, J.L. Sonnenberg, M. Hada, M. Ehara, K. Toyota, R. Fukuda, J. Hasegawa, M. Ishida, T. Nakajima, Y. Honda, O. Kitao, H. Nakai, T. Vreven, J.A. Montgomery, Jr., J.E. Peralta, F. Ogliaro, M. Bearpark, J.J. Heyd, E. Brothers, K.N. Kudin, V.N. Staroverov, R. Kobayashi, J. Normand, K. Raghavachari, A. Rendell, J.C. Burant, S.S. Iyengar, J. Tomasi, M. Cossi, N. Rega, J.M. Millam, M. Klene, J.E. Knox, J.B. Cross, V. Bakken, C. Adamo, J. Jaramillo, R. Gomperts, R.E. Stratmann, O. Yazyev, A. J. Austin, R. Cammi, C. Pomelli, J.W. Ochterski, R.L. Martin, K. Morokuma, V.G. Zakrzewski, G.A. Voth, P. Salvador, J.J. Dannenberg, S. Dapprich, A.D. Daniels, Ö. Farkas,

J.B. Foresman, J.V. Ortiz, J. Cioslowski, D.J. Fox, *Gaussian 09, Revision A.02* (Gaussian, Inc., Wallingford, 2009)

3. H.-J. Werner, P.J. Knowles, R. Lindh, F.R. Manby, M. Schütz, and others MOLPRO, version 2009.1, a package of *ab initio* programs, see http://www.molpro.net

4. G. Maroulis, A systematic study of basis set, electron correlation, and geometry effects on the electric multipole moments, polarizability, and hyperpolarizability of HCl. J. Chem. Phys. **108**(13), 5432–5448 (1998)

5. B. Linder, R.A. Kromhout, Van der Waals induced dipoles. J. Chem. Phys. **84**(5), 2753–2760 (1986)

6. P.W. Fowler, Dispersion dipoles, quadrupoles and electric-field gradients. Chem. Phys. **143**, 447–457 (1990)

7. X. Li, M.H. Champagne, K.L.C. Hunt, Long-range, collision-induced dipoles of Td-D∞h molecule pairs: theory and numerical results for CH4 or CF4 interacting with H2, N2, CO2, or CS2. J. Chem. Phys. **109**(19), 8416–8425 (1998)

8. J.E. Bohr, K.L.C. Hunt, Dipoles induced by van der Waals interactions during collisions of atoms with heteroatoms or with centrosymmetric linear molecules. J. Chem. Phys. **86**(10), 5441–5448 (1987)

9. M.H. Champagne, X. Li, K.L.C. Hunt, Nonadditive three-body polarizabilities of molecules interacting at long range: theory and numerical results for the inert gases, H2, N2, CO2, and CH4. J. Chem. Phys. **112**(4), 1893–1906 (2000)

10. A. Unsöld, Quantentheorie des Wasserstoffmolekülions und der Born-Landéschen Abstoβungskräfte. Z. Physik. **43**, 563–574 (1927)

11. B.M. Smirnov, in *Asymptotic Methods in Atomic Collisions* (Atompress, Moscow, 1973, in Russian)

12. B.M. Smirnov, E.E. Nikitin, in *Atomic and Molecular Processes* (Nauka, Moscow, 1988, in Russian)

13. N. Zvereva-Loëte, YuN Kalugina, V. Boudon, M.A. Buldakov, V.N. Cherepanov, Dipole moment surface of the van der Waals complex CH4–N2. J. Chem. Phys. **133**, 184302 (2010)

14. D. Xenides, A. Hantzis, G. Maroulis, Comparison of high-level post-Hartree–Fock and DFT methods on the calculation of interaction-induced electric properties of Kr–He. Chem. Phys. **382**, 80–87 (2011)

15. A. Haskopoulos, G. Maroulis, Interaction dipole moment in Rg–Xe (Rg=He, Ne, Ar, and Kr) heterodiatoms from conventional *ab initio* and density functional theory calculations. J. Math. Chem. **40**(3), 233–242 (2006)

16. L. Galatry, T. Gharbi, The long-range dipole moment of two interacting spherical systems. Chem. Phys. Lett. **75**(3), 427–433 (1980)

17. D.P. Craig, Elementaty derivation of long-range moments of two coupled centrosymmetric systems. Chem. Phys. Lett. **80**(1), 14–17 (1981)

18. W.B. Brown, D.M. Whisnant, Interatomic dispersion dipole. Mol. Phys. **26**(5), 1105–1119 (1973)

19. D.M. Whisnant, W.B. Brown, Interatomic dispersion dipole. Mol. Phys. **25**(6), 1385–1403 (1973)

20. J. Mahanty, C.K. Majumdar, Exchange-induced dipole moments in atom pairs. Phys. Rev. A **26**(5), 2334–2337 (1982)

21. P. Karamanis, G. Maroulis, How Important are high-level *ab initio* treatments for the interaction dipole moment and polarizability of HeNe? Comput. Lett. (CoLe) **1**(3), 117 (2005)

22. J. Goodisman, Dipole-moment function for diatomic molecules. J. Chem. Phys. **38**(11), 2597–2599 (1963)

23. M.A. Buldakov, V.N. Cherepanov, The semiempirical dipole moment functions of the molecules HX (X=F, Cl, Br, I, O), CO and NO. J. Phys. B: At. Mol. Opt. Phys. **37**(19), 3973–3986 (2004)

24. R.J. Hinde, Interaction-induced dipole moment of Ar-H_2 dimer: dependence on the H_2 bond length. J. Chem. Phys. **124**, 154309 (2006)

25. V.P. Bulychev, K.M. Bulanin, M.O. Bulanin, Theoretical study of the Spectral and structural parameters of van der Waals complexes of the Li^+ cation with the H_2, D_2, and T_2 isotopomers of the hydrogen molecule. Opt. Spectrosc. **96**(2), 205–216 (2004)

26. F. Wang, F.R.W. McCourt, R.J. Le Roy, Dipole moment surfaces and the mid- and far-IR spectra of N_2-Ar. J. Chem. Phys. **113**(1), 98–106 (2000)

27. F. Wang, F.R.W. McCourt, R.J. Le Roy, Use of simulated infrared spectra to test N_2-Ar pair potentials and dipole moment surfaces. Molec. Phys. **88**(3), 821–840 (1996)

28. A.L. Cooksy, M.J. Elrod, R.J. Saykally, W. Klemperer, Dipole moment analysis of excited van der Waals vibrational states of $ArH^{35}Cl$. J. Chem. Phys. **99**(5), 3200–3204 (1993)

29. A.G. Ayllon, J. Santamaria, S. Miller, J. Tennyson, Calculated spectra for N_2-Ar van der Waals complex. Molec. Phys. **71**(5), 1043–1054 (1990)

30. W. Meyer, L. Frommhold, Collision-induced rototranslational spectra of H_2-He from an accurate *ab initio* dipole moment surface. Phys. Rev. A **34**(4), 2771–2779 (1986)

31. Y.N. Kalugina, S.E. Lokshtanov, V.N. Cherepanov, A.A. Vigasin, *Ab initio* 3D potential energy and dipole moment surfaces for the CH4–Ar complex: collision-induced intensity and dimer content. J. Chem. Phys. **144**(5), 054304 (2016)

32. L. Frommhold, in *Collision-Induced Absorption in Gases* (Cambridge University Press, Cambridge, 1993)

33. Weakly interacting molecular pairs: Unconventional absorbers of radiation in the atmosphere. NATO Science Series. IV. *Earth and Environmental Sciences*, vol. 27, eds. by C. Camy-Preyt, A. Vigasin; A. Vigasin and Z. Slanina, *Molecular Complexes in Earth's Planetary, Cometary and Interstellar Atmospheres* (World Scientific Publishing, 1998)

34. M. Afshari, M. Dehghany, J. Norooz Oliaee, N. Moazzen-Ahmadi, Infrared spectra of the OCS–N_2O complex and observation of a new isomer. Chem. Phys. Lett. **489**(1–3), 30 (2010)

35. A. Baranowska, B. Fernandez, A. Rizzo, B. Jansik, The CO–Ne van der Waals complex: *ab initio* intermolecular potential energy, interaction induced electric dipole moment and polarizability surfaces, and second viral coefficients. Phys. Chem. Chem. Phys. **11**, 9871 (2009)

36. T. Bancewicz, G. Maroulis, Rotationally adapted studies of *ab initio*–computed collision-induced hyperpolarizabilities: The H_2-Ar pair. Phys. Rev A **79**, 042704 (2009)

37. X. Li, K.L.C. Hunt, F. Wang, M. Abel, L. Frommhold, Collision-induced infrared absorption by molecular hydrogen pairs at thousands of Kelvin. Int. J. Spectrosc. **2010**, 371201 (2009)

38. K. Didriche, C. Lauzin, P. Macko, M. Herman, W.J. Lafferty, Observation of the C_2H_2–N_2O van der Waals complex in the overtone range using CW-CRDS. Chem. Phys. Lett. **469**(1–3), 35 (2009)

39. P. Macko, C. Lauzin, M. Herman, High resolution spectroscopy of the 2CH band in the $^{12}C_2H_2$–Ar van der Waals complex. Chem. Phys. Lett. **445**(4–6), 113 (2007)

40. Q. Wen, W. Jäger, Microwave and *ab initio* studies of the Xe–CH_4 van der Waals complex. J. Chem. Phys. **124**(1), 014301 (2006)

41. W.C. Topic, W. Jäger, The weakly bound He–HCCCN complex: high-resolution microwave spectra and intermolecular potential-energy surface. J. Chem. Phys. **123**(6), 064303 (2005)

42. W.M. Fawzy, G. Kerenskaya, M.C. Heaven, Experimental detection and theoretical characterization of the H_2–NH(X) van der Waals complex. J. Chem. Phys. **122**(14), 144318 (2005)

43. Y. Liu, W. Jäger, Microwave and *ab initio* studies of rare gas–methane van der Waals complexes. J. Chem. Phys. **120**(19), 9047 (2004)

44. Y. Liu, W. Jäger, Microwave investigation of the CO-CH_4 van der Waals complex. J. Chem. Phys. **121**(13), 6240 (2004)

45. F. Raulin, D. Mourey, G. Toupance, Organic syntheses from CH_4-N_2 atmospheres: implications for Titan. Orig. Life. **12**(3), 267–279 (1982)

46. A. Coustenis, F.W. Taylor, in *Titan: Exploring an Earthlike World* (World Scientific Publishing Co. Pte. Ltd., 2008)

47. R. Courtin, The spectrum of titan in the far-infrared and microwave regions. ICARUS **51**(3), 466–475 (1982)
48. R. Courtin, Pressure-induced absorption coefficients for radiative transfer calculations in Titan's atmosphere. ICARUS **75**(2), 245–254 (1988)
49. O.B. Toon, C.P. McKay, R. Courtin, T.P. Ackerman, Methane rain on Titan. ICARUS **75** (2), 255–284 (1988)
50. G.F. Lindal, G.E. Wood, H.B. Hotz, D.N. Sweetnam, V.R. Eshleman, G.L. Tyler, The atmosphere of Titan: an analysis of the Voyager 1 radio occultation measurements. ICARUS **53**(2), 348–363 (1983)
51. W. Reid Thompson, J.A. Zollweg, D.H. Gabis, Vapor-liquid equilibrium thermodynamics of N_2+ CH_4: model and Titan applications. ICARUS **97**(2), 187–199 (1992)
52. A. Borysow, C. Tang, Far infrared CIA spectra of N_2-CH_4 pairs for modeling of Titan's atmosphere. ICARUS **105**(1), 175–183 (1993)
53. R. Courtin, D. Gautier, C.P. McKay, Titan's thermal emission spectrum: reanalysis of the voyager infrared measurements. ICARUS **114**(1), 144–162 (1995)
54. C.P. McKay, Elemental composition, solubility, and optical properties of Titan's organic haze Planet. Space Sci. **44**(8), 741–747 (1996)
55. R.E. Samuelson, N.R. Nath, A. Borysow, Gaseous abundances and methane supersaturation in Titan's troposphere planet. Space Sci. **45**(8), 959–980 (1997)
56. F.M. Flasar, R.K. Achterberg, B.J. Conrath, P.J. Gierasch, V.G. Kunde, C.A. Nixon, G.L. Bjoraker, D.E. Jennings, P.N. Romani, A.A. Simon-Miller, B. Bezard, A. Coustenis, P.G. J. Irwin, N.A. Teanby, J. Brasunas, J.C. Pearl, M.E. Segura, R.C. Carlson, A. Mamoutkine, P.J. Schinder, A. Barucci, R. Courtin, T. Fouchet, D. Gautier, E. Lellouch, A. Marten, R. Prange, S. Vinatier, D.F. Strobel, S.B. Calcutt, P.L. Read, F.W. Taylor, N. Bowles, R.E. Samuelson, G.S. Orton, L.J. Spilker, T.C. Owen, J.R. Spencer, M.R. Showalter, C. Ferrari, M.M. Abbas, F. Raulin, S. Edgington, P. Ade, E.H. Wishnow, Titan's atmospheric temperatures, winds, and composition. Science **308**, 975–978 (2005)
57. S.J. Kim, T.R. Geballe, K.S. Noll, R. Courtin, Clouds, haze, and CH_4, CH_3D, HCN, and C_2H_2 in the atmosphere of Titan probed via 3 μm spectroscopy. ICARUS **173**(2), 522–532 (2005)
58. A. Coustenis, R.K. Achterberg, B.J. Conrath, D.E. Jennings, A. Marten, D. Gautier, C.A. Nixon, F.M. Flasar, N.A. Teanby, B. Bzard, R.E. Samuelson, R.C. Carlson, E. Lellouch, G. L. Bjoraker, P.N. Romani, F.W. Taylor, P.G.J. Irwin, T. Fouchet, A. Hubert, G.S. Orton, V. G. Kunde, S. Vinatier, J. Mondellini, M.M. Abbas, R. Courtin, The composition of Titan's stratosphere from Cassini/CIRS mid-infrared spectra. ICARUS **189**(1), 35–62 (2007)
59. H. Seo, S.J. Kim, J.H. Kim, T.R. Geballe, R. Courtin, L.R. Brown, Titan at 3 microns: newly identified spectral features and an improved analysis of haze opacity. ICARUS **199** (2), 449–457 (2009)
60. D.E. Jennings, F.M. Flasar, V.G. Kunde, R.E. Samuelson, J.C. Pearl, C.A. Nixon, R.C. Carlson, A.A. Mamoutkine, J.C. Brasunas, E. Guandique, R.K. Achterberg, G.L. Bjoraker, P.N. Romani, M.E. Segura, S.A. Albright, M.H. Elliott, J.S. Tingley, S. Calcutt, A. Coustenis, R. Courtin, Titan's surface brightness temperatures. Astrophys. J. **691**(1), L103–L105 (2009)
61. H. Schindler, R. Vogelsang, V. Staemmler, M.A. Siddiqi, P. Svejda, *Ab initio* intermolecular potentials of methane, nitrogen methane + nitrogen and their use in Monte Carlo simulations of fluids and fluid mixtures. Mol. Phys. **80**(6), 1413 (1993)
62. M. Shadman, S. Yeganegi, F. Ziaie, *Ab initio* interaction potential of methane and nitrogen. Chem. Phys. Lett. **467**, 237 (2009)
63. Y.N. Kalugina, V.N. Cherepanov, M.A. Buldakov, N. Zvereva-Loëte, V. Boudon, Theoretical investigation of the potential energy surface of the van der Waals complex CH_4–N_2. J. Chem. Phys. **131**, 134304 (2009)
64. M.A. Buldakov, V.N. Cherepanov, YuN Kalugina, N. Zvereva-Loëte, V. Boudon, Static polarizability surfaces of the van der Waals complex CH_4–N_2. J. Chem. Phys. **132**(16), 164304 (2010)

65. M. Buser, L. Frommhold, M. Gustafsson, M. Moraldi, M.H. Champagne, K.L.C. Hunt, Far-infrared absorption by collisionally interacting nitrogen and methane molecules. J. Chem. Phys. **121**(6), 2617 (2004)

66. M. Buser, L. Frommhold, Infrared absorption by collisional CH_4-X pairs, with X=He, H_2, or N_2. J. Chem. Phys. **122**(2), 024301 (2005)

67. I.R. Dagg, A. Anderson, S. Yan, W. Smith, C.G. Joslin, L.A.A. Read, Collision-induced absorption in gaseous mixtures of nitrogen and methane. Can. J. Phys. **64**(11), 1467–1474 (1986)

68. G. Birnbaum, A. Borysow, A. Buechele, Collision-induced absorption in mixtures of symmetrical linear and tetrahedral molecules: methane-nitrogen. J. Chem. Phys. **99**(5), 3234 (1993)

69. D.J. Margoliash, W.J. Meath, Pseudospectral dipole oscillator strength distributions and some related two body interactions coefficients for H, He, Lim N, O, H_2, N_2, O_2, NO, N_2, H_2O, NH_3 and CH_4. J. Chem. Phys. **68**(4), 1426 (1978)

70. G. Maroulis, Accurate electric multipole moment, static polarizability and hyperpolarizability derivatives for N_2. J. Chem. Phys. **118**(6), 2673 (2003)

71. G. Maroulis, Electric dipole hyperpolarizability and quadrupole polarizability of methane from finite-field coupled cluster and fourth-order many-body perturbation theory calculations. Chem. Phys. Lett. **226**, 420 (1994)

72. G. Maroulis, Dipole–quadrupole and dipole–octopole polarizability for CH_4 and CF_4. J. Chem. Phys. **105**(18), 8467 (1996)

73. C. Huiszoon, *Ab initio* calculations of multipole moments, polarizabilities and isotropic long-range coefficients for dimethylether, methanol, methane, and water. Mol. Phys. **58**, 865 (1986)

74. M. Hashimoto, T. Isobe, CNDO/2 calculation of the valence electron contribution to the intermolecular potential of some ground state closed shell molecules. Bull. Chem. Soc. Jpn. **46**(8), 2581–2582 (1973)

75. M. Hashimoto, T. Isobe, The INDO and CNDO/2 SCF LCAO MO calculation of intermolecular forces and their pairwise additivity. Bull. Chem. Soc. Jpn. **47**(1), 40–44 (1974)

76. K. Suzuki, K. Iguchi, The intermolecular potential of the ethylene dimer. J. Chem. Phys. **75** (9), 779–784 (1978)

77. BKh Khalbaev, I.A. Misurkin, Intermolecular interactions in the ethylene dimer according to perturbation theory in the CNDO/2 approximation with a new formula for the resonance integral. Theor. Exp. Chem. **20**(4), 365–372 (1984)

78. BKh Khalbaev, I.A. Misurkin, Investigation of the intermolecular interaction in the ethylene dimer by a modified CNDO method. Theor. Exp. Chem. **21**(5), 505–512 (1985)

79. V. Brenner, Ph Millie, Intermolecular interactions: basis set and intermolecular correlation effects on semiempirical methods. Application to $(C_2H_2)_2$, $(C_2H_2)_3$ and $(C_2H_4)_2$. Z. Phys. D **30**(4), 327–340 (1994)

80. P.E.S. Wormer, A. van der Avoird, *Ab initio* valence-bond calculations of the van der Waals interactions between π systems: the ethylene dimer. J. Chem. Phys. **62**(8), 3326–3339 (1975)

81. T. Wasiutynski, A. van der Avoird, R.M. Berns, Lattice dynamics of the ethylene crystal with interaction potentials from *ab initio* calculations. J. Chem. Phys. **69**(12), 5288–5300 (1978)

82. T. Luty, A. van der Avoird, R.M. Berns, T. Wasiutynski, Dynamical and optical properties of the ethylene crystal: self-consistent phonon calculations using an *ab initio* intermolecular potential. J. Chem. Phys. **75**(3), 1451–1458 (1981)

83. E.J.P. Malar, A.K. Chandra, Intermolecular potentials in the dimer, the excimers, and the dimer ions of ethylene. J. Phys. Chem. **85**(15), 2190–2194 (1981)

84. K. Suzuki, K. Iguchi, *Ab initio* intermolecular potential of the ethylene dimer. J. Chem. Phys. **77**(9), 4594–4603 (1982)

85. I.L. Alberts, T.W. Rowlands, N.C. Handy, Stationary points on the potential energy surfaces of $(C_2H_2)_2$, $(C_2H_2)_3$ and $(C_2H_4)_2$. J. Chem. Phys. **88**(6), 3811–3816 (1988)

86. S. Tsuzuki, K. Tanabe, Nonbonding interaction potential of ethylene dimer obtained from *ab initio* molecular orbital calculations: Prediction of a D_{2d} structure. J. Phys. Chem. **96**(26), 10804–10808 (1992)
87. E. Rytter, D.M. Gruen, Infrared spectra of matrix isolated and solid ethylene. Formation of ethylene dimers. Spectrochim. Acta A **35**(3), 199–207 (1979)
88. M.C. Chan, P.A. Block, R.E. Miller, Structure of the ethylene dimer from rotationally resolved near-infrared spectroscopy: a quadruple hydrogen bond. J. Chem. Phys. **102**(10), 3993–3999 (1995)
89. S. Tsuzuki, T. Uchimaru, K. Tanabe, Intermolecular interaction potentials of methane and ethylene dimers calculated with the Møller-Plesset, coupled cluster and density functional methods. Chem. Phys. Lett. **287**(1–2), 202–208 (1998)
90. S. Tsuzuki, T. Uchimaru, K. Matsumura, M. Mikami, K. Tanabe, Effects of the higher electron correlation correction on the calculated intermolecular interaction energies of benzene and naphthalene dimers: comparison between MP2 and CCSD(T) calculations. Chem. Phys. Lett. **319**(5–6), 547–554 (2000)
91. S. Tsuzuki, T. Uchimaru, M. Mikami, K. Tanabe, New medium-size basis sets to evaluate the dispersion interaction of hydrocarbon molecules. J. Phys. Chem. A **102**(12), 2091–2094 (1998)
92. P. Jurečka, J. Šponer, J. Černý, P. Hobza, Benchmark database of accurate (MP2 and CCSD (T) complete basis set limit) interaction energies of small model complexes, DNA base pairs, and amino acid pairs. Phys. Chem. Chem. Phys. **8**(17), 1985–1993 (2006)
93. J. Antony, S. Grimme, Is spin-component scaled second-order Møller-Plesset perturbation theory an appropriate method for the study of noncovalent interactions in molecules? J. Phys. Chem. A **111**(22), 4862–4868 (2007)
94. R.A. King, On the accuracy of spin-component-scaled perturbation theory (SCS-MP2) for the potential energy surface of the ethylene dimer. Mol. Phys. **107**(8–12), 789–795 (2009)
95. F. Mulder, C. Huiszoon, The dimer interaction and lattice energy of ethylene and pyrazine in the multipole expansion; a comparison with atom-atom potentials. Mol. Phys. **34**(5), 1215–1235 (1977)
96. F. Mulder, M. van Hemert, P.E.S. Wormer, A. van der Avoird, *Ab initio* studies of long range interactions between ethylene molecules in the multipole expansion. Theor. Chim. Acta **46**(1), 39–62 (1977)
97. P.E.S. Wormer, F. Mulder, A. van der Avoird, Quantum theoretical calculations of van der Waals interactions between molecules. Anisotropic long range interactions. Int. J. Quant. Chem. **11**(6), 959–970 (1977)
98. P. Coulon, R. Luyckx, H.N.W. Lekkerkerker, Approximate calculation of the dynamic polarizabilities and dispersion interaction for ethylene molecules. J. Chem. Soc., Faraday Trans. 2 **77**(1), 201–207 (1981)
99. R. Ahlrichs, S. Brode, U. Buck, M. DeKieviet, C. Lauenstein, A. Rudolph, B. Schmidt, The structure of C_2H_4 clusters from theoretical interaction potentials and vibrational predissociation data. Z. Phys. D **15**(4), 341–351 (1990)
100. YuN Kalugina, V.N. Cherepanov, M.A. Buldakov, N. Zvereva-Loëte, Vincent Boudon, Theoretical investigation of the ethylene dimer: interaction energy and dipole moment. J. Comput. Chem. **33**(3), 319–330 (2012)
101. C.G. Gray, K.E. Gubbins, I.R. Dagg, L.A.A. Read, Determination of the quadrupole moment tensor of ethylene by collision-induced absorption. Chem. Phys. Lett. **73**(2), 278–282 (1980)
102. I.R. Dagg, L.A.A. Read, B. Andrews, Collision-induced absorption in ethylene in the microwave and far-infrared regions. Can. J. Phys. **59**(1), 57–65 (1981)
103. I.R. Dagg, L.A.A. Read, B. Andrews, Collision-induced absorption in the far infrared region in ethylene—rare gas mixtures. Can. J. Phys. **60**(10), 1431–1441 (1982)
104. W.C. Pringle, R.C. Cohen, S.M. Jacobs, Analysis of collision induced far infrared spectrum of ethylene. Mol. Phys. **62**(3), 661–668 (1987)
105. A. Kumar, B.L. Jhanwar, W. Meath, Can. J. Chem. **85**, 724 (2007)

106. A.D. McLean, M. Yoshimine, J. Chem. Phys. **47**, 1927 (1967)
107. P. Karamanis, G. Maroulis, Electric quadrupole and hexadecapole moments for $X_2C=CX_2$, X=H, F, Cl, Br, and I. Int. J. Quant. Chem. **90**, 483–490 (2002)
108. J.L. Duncan, I.J. Wright, D. van Leberghe, Ground state rotational constants of H_2CCD_2 and C_2D_4 and geometry of ethylene. J. Mol. Spectrosc. **42**, 463–477 (1972)
109. G. Maroulis, A study of basis set and electron correlation effects in the *ab initio* calculation of the electric dipole hyperpolarizability of ethene ($H_2C=CH_2$). J. Chem. Phys. **97**(6), 4188–4194 (1992)

Chapter 4
Interaction-induced Polarizability

The methods for computation of molecular polarizability are implemented now in many well-known modern quantum chemical codes. Some of their features will be discussed in the next Section. And here, we point out the experimental methods used usually to determine the polarizabilities of atoms, molecules or their complexes. For this purpose, some measured macroscopic properties are used. As a rule, these properties are some functions of the electromagnetic field which are changing during the interaction of molecules or atoms with a field. Herewith, the accuracy of determination of the polarizability depends on the observation methods and the aggregate state of an environment. The most accurate results are obtained for incompact gas media.

The most known method to evaluate the polarizability is based on measurements of the molecular refractivity of any gas. This method allows determining with high accuracy the average polarizability of molecules [1]. So, for nitrogen and oxygen molecules the average polarizabilities were measured with the accuracy up to 0.01–0.03 % [2–6].

Another widespread experimental method uses the changes of parameters for Rayleigh light scattering. In this way, both invariants of the polarizability tensor (the average polarizability and the polarizability anisotropy) from the light scattering intensity can be found. However, in practice, the anisotropy polarizability is preferred for determination by this method. This is because of the fact that for this case the average polarizability is determined less accurate in comparison with the refraction method [7, 8]. At the same time, the real accuracy of the polarizability anisotropy obtained by the use the polarization characteristics of the light scattering doesn't exceed of 3–5 %.

The Kerr effect is also often used to determine the polarizabilities of molecules [1, 9, 10]. In the general case, the Kerr constant has a complicated relation with the polarizability anisotropy, dipole moment and hyperpolarizability of a molecule. As a result, the problem of a separation of the contributions to the Kerr constant from these values is appeared. The separation difficulties lead to the increase of errors of the method (10–20 %). Only for the gas media composed of non-polar molecules

V.N. Cherepanov et al., *Interaction-induced Electric Properties of van der Waals Complexes*, SpringerBriefs in Electrical and Magnetic Properties of Atoms, Molecules, and Clusters, DOI 10.1007/978-3-319-49032-8_4

the dependence of the Kerr constant on electric properties of a molecule is significantly simplified and the polarizability anisotropy can be directly determined from the Kerr constant.

The experimental methods for polarizability determination based on the Stark effect use the displacements and splitting of rotational energy levels of small molecules under the influence of an external electric field [11–14]. The usual accuracy of these methods is in the range of 3–10 %.

Finally, let us mention the experimental methods that use other physical effects to measure molecular polarizability. These methods use the birefringence effects [15] in any magnetic field (Cotton–Mouton effect) and flow (dynamic optical effect of Maxwell), the acoustic birefringence effect, absorption spectra induced by the electric field [16] and so on. It should be noted that last group of methods have the greater errors compared to the methods discussed above.

The feature of the all considered experimental methods is that they allow us to define the values of the molecular polarizability only at the equilibrium position of molecular nuclei. To obtain the dependencies of molecular polarizabilities on the mutual location of nuclei in a molecule, the Raman effect can be used The line intensities of Raman spectra depend on the values of polarizability derivatives with respect to the nuclei displacements. The first works to define the polarizability derivatives of molecules have appeared immediately after the creation of the theory of Raman light scattering (Placzek theory of polarizability) [17]. However, the experimental technique of "pre-laser" period could not obtain the high-quality results. Some experimental results of this period are summarized in [18]. Currently, these data have only a historical interest. Now, laser technologies allow to increase the measurement accuracy and, as a result, significantly improve and revise the "pre-laser" data. Nevertheless, up to day the experimental data on the polarizability derivatives of molecules are fragmentary and do not give the impression of systematic studies of the polarizability of molecules as a function of the nuclei coordinates, even for diatomic molecules [19–39].

Note also, that U. Hohm has recently compiled the static mean dipole-dipole polarizability evaluated from gas phase measurements for 174 molecules [40].

4.1 Interaction-induced Polarizability Theory

4.1.1 Ab Initio Calculation Features

The polarizability is the second derivative of the interaction energy by the external field F_α^0. The calculation formulas can be also obtained using the finite-difference method like for the dipole moment. As a result, for example, the 3-point finite difference approximation (with errors of order $(F_\alpha^0)^2$) gives

$$\alpha_{\alpha\alpha} = -\frac{E(F_\alpha^0, F_\alpha^0) - 2E(0,0) + E(-F_\alpha^0, -F_\alpha)}{2(F_\alpha^0)^2},$$

$$\alpha_{\alpha\beta} = -\frac{E(F_\alpha^0, F_\beta^0) - E(F_\alpha^0, -F_\beta^0) - E(-F_\alpha^0, F_\beta^0) + E(-F_\alpha^0, -F_\beta^0)}{4F_\alpha^0 F_\beta^0}, \quad \alpha \neq \beta.$$

(4.1.1)

Also, the more accurate formula proposed by Maroulis [41] allows to eliminate the contribution of higher-order terms to the polarizability:

$$\alpha_{\alpha\alpha} = \frac{1024 S_\alpha(F_\alpha^0) - 80 S_\alpha(2F_\alpha^0) + S_\alpha(4F_\alpha^0)}{360(F_\alpha^0)^2}, \qquad (4.1.2)$$

where

$$S_\alpha(F_\alpha^0) = \frac{E(-F_\alpha^0) + E(F_\alpha^0) - 2E(0)}{2}.$$

It should be pointed out that both for the dipole moment and polarizability of interacting molecules (complexes), one should account for the BSSE correction. This means that the single point energy calculations with different external fields should be carried out with the BSSE correction. The BSSE correction depends on the basis set employed and the system under consideration. For some cases the BSSE correction has negligible effect on electric properties, and in this case it could be neglected. However, the more correct way is to take into account the BSSE correction. The choice of the applied homogeneous field should be done very carefully. For this purpose one should carry out a series of calculations with different amplitudes of the external field F_α^0. From these calculations the range of the amplitudes of the field can be found where the property under the investigation doesn't change significantly with the change of the amplitude F_α^0. The field only from this range can be used for the further calculation. Sometimes for different properties the different amplitudes of the external field should be applied.

4.1.2 Long-Range Approximation

It should be pointed out that the methods of classical electrodynamics accounting for the induction and dispersion effects give a physically correct analytical description of the polarizability surface for interacting atomic-molecular systems at large intermolecular separations [42–54]. In this way, in the framework of the long-range approximation [1, 55], when the interacting species are considered as point objects with their anisotropic electric properties, the electric polarizability $\alpha_{\alpha\beta}^{AB}$ of two interacting systems may be written in the form

$$\alpha_{\alpha\beta}^{AB} = \alpha_{\alpha\beta}^{A} + \alpha_{\alpha\beta}^{B} + \Delta\alpha_{\alpha\beta}^{AB}, \tag{4.1.3}$$

where $\alpha_{\alpha\beta}^{A}$ and $\alpha_{\alpha\beta}^{B}$ are the dipole polarizabilities of the atomic (or molecular) systems A and B, and $\Delta\alpha_{\alpha\beta}^{AB}$ is the interaction polarizability. Here, the interaction polarizability $\Delta\alpha_{\alpha\beta}^{AB}$ can be written as follows

$$\Delta\alpha_{\alpha\beta}^{AB} = \alpha_{\alpha\beta}^{ind} + \alpha_{\alpha\beta}^{disp} + \alpha_{\alpha\beta}^{exch}, \tag{4.1.4}$$

where $\alpha_{\alpha\beta}^{ind} = \alpha_{\alpha\beta}^{ind,A} + \alpha_{\alpha\beta}^{ind,B}$, $\alpha_{\alpha\beta}^{disp} = \alpha_{\alpha\beta}^{disp,A} + \alpha_{\alpha\beta}^{disp,B}$ and $\alpha_{\alpha\beta}^{exch}$ are the induction, dispersion and exchange contributions to the polarizability of interacting systems A and B. Evidently, when the interacting systems are well separated, the interaction polarizability $\Delta\alpha_{\alpha\beta}^{AB}$ is determined entirely by a well-known induction and dispersion contributions $\alpha_{\alpha\beta}^{ind}$ and $\alpha_{\alpha\beta}^{disp}$ [43–45, 48, 56–58]. At shorter range, when the charge distributions of interacting systems overlap, the exchange effects appear and the additional contribution $\alpha_{\alpha\beta}^{exch}$ begins to be important [59].

The induction contribution to the polarizability $\alpha_{\alpha\beta}^{ind,A}$ of the A molecule is determined in accordance with (2.4.2) at $F_{\alpha}^{0} = 0$. Then, using analogous scheme like for calculations of the induction dipole moment, the following expression can be obtained for the induction polarizability of the molecule A:

$$
\begin{aligned}
\alpha_{\alpha\beta}^{ind,A} =\ & \alpha_{\alpha\gamma}^{A} T_{\gamma\delta} \alpha_{\delta\beta}^{B} + \beta_{\alpha\beta\gamma}^{A} T_{\gamma\delta} \mu_{\delta}^{B} + \frac{1}{3} \alpha_{\alpha\gamma}^{A} T_{\gamma\delta\varepsilon} A_{\beta,\delta\varepsilon}^{B} - \frac{1}{3} A_{\alpha,\gamma\delta}^{A} T_{\gamma\delta\varepsilon} \alpha_{\varepsilon\beta}^{B} + \frac{1}{3} \beta_{\alpha\beta\gamma}^{A} T_{\gamma\delta\varepsilon} \Theta_{\delta\varepsilon}^{B} \\
& - \frac{1}{3} B_{\alpha\beta,\gamma\delta}^{A} T_{\gamma\delta\varepsilon} \mu_{\varepsilon}^{B} + \frac{1}{15} \alpha_{\alpha\gamma}^{A} T_{\gamma\delta\varepsilon\varphi} E_{\beta,\delta\varepsilon\varphi}^{B} + \frac{1}{15} E_{\alpha,\gamma\delta\varepsilon}^{A} T_{\gamma\delta\varepsilon\varphi} \alpha_{\varphi\beta}^{B} - \frac{1}{9} A_{\alpha,\gamma\delta}^{A} T_{\gamma\delta\varepsilon\varphi} A_{\beta,\varepsilon\varphi}^{B} \\
& + \frac{1}{15} \beta_{\alpha\beta\gamma}^{A} T_{\gamma\delta\varepsilon\varphi} \Omega_{\delta\varepsilon\varphi}^{B} - \frac{1}{9} B_{\alpha\beta,\gamma\delta}^{A} T_{\gamma\delta\varepsilon\varphi} \Theta_{\varepsilon\varphi}^{B} + \frac{1}{15} M_{\alpha\beta,\gamma\delta}^{A} T_{\gamma\delta\varepsilon\varphi} \mu_{\varphi}^{B} + \frac{1}{105} \alpha_{\alpha\gamma}^{A} T_{\gamma\delta\varepsilon\varphi\nu} D_{\beta,\delta\varepsilon\varphi\nu}^{B} \\
& - \frac{1}{105} D_{\alpha,\gamma\delta\varepsilon\varphi}^{A} T_{\gamma\delta\varepsilon\varphi\nu} \alpha_{\nu\beta}^{B} - \frac{1}{45} A_{\alpha,\gamma\delta}^{A} T_{\gamma\delta\varepsilon\varphi\nu} E_{\beta,\varepsilon\varphi\nu}^{A} + \frac{1}{45} E_{\alpha,\gamma\delta\varepsilon}^{A} T_{\gamma\delta\varepsilon\varphi\nu} A_{\beta,\varphi\nu}^{A} + \frac{1}{45} M_{\alpha\beta,\gamma\delta}^{A} T_{\gamma\delta\varepsilon\varphi\nu} \Theta_{\varphi\nu}^{B} \\
& - \frac{1}{105} G_{\alpha\beta,\gamma\delta\varepsilon\varphi}^{A} T_{\gamma\delta\varepsilon\varphi\nu} \mu_{\nu}^{B} + \frac{1}{105} \beta_{\alpha\beta\gamma}^{A} T_{\gamma\delta\varepsilon\varphi\nu} \Phi_{\delta\varepsilon\varphi\nu}^{B} - \frac{1}{45} B_{\alpha\beta,\gamma\delta}^{A} T_{\gamma\delta\varepsilon\varphi\nu} \Omega_{\varepsilon\varphi\nu}^{B} + \frac{1}{315} A_{\alpha,\beta\gamma}^{A} \Phi_{\delta\varepsilon\varphi\nu}^{B} T_{\beta\gamma\delta\varepsilon\varphi\nu} \\
& + \alpha_{\alpha\gamma}^{A} T_{\gamma\delta} \alpha_{\delta\varepsilon}^{B} T_{\varepsilon\varphi} \alpha_{\varphi\beta}^{A} + \alpha_{\alpha\gamma}^{A} T_{\gamma\delta} \beta_{\delta\beta\varepsilon}^{B} T_{\varepsilon\varphi} \mu_{\varphi}^{A} + \beta_{\alpha\gamma\delta}^{A} T_{\gamma\varepsilon} \mu_{\varepsilon}^{B} T_{\delta\varphi} \alpha_{\varphi\beta}^{A} + \beta_{\alpha\gamma\delta}^{A} T_{\gamma\delta} \alpha_{\delta\varepsilon}^{B} T_{\varepsilon\varphi} \mu_{\varphi}^{A} + \cdots.
\end{aligned}
$$

$$\tag{4.1.5}$$

Here $T \equiv T^{AB}$.

The dispersion contribution to the polarizability of the interacting molecules can be found by calculation of the second derivative of the interaction energy

$$\alpha_{\alpha\beta}^{disp} = -\left.\frac{\partial E_{disp}^{AB}}{\partial F_{\alpha}^{0} F_{\beta}^{0}}\right|_{F^{0}=0} = \alpha_{\alpha\beta}^{disp,A} + \alpha_{\alpha\beta}^{disp,B} \tag{4.1.6}$$

where, restricting ourselves to the leading term $\sim R^{-6}$, the expression for the dispersion contribution takes the form (see (2.3.5), [58])

$$\alpha_{\alpha\beta}^{disp,A} = \frac{1}{2\pi}\int\limits_{0}^{\infty} d\omega\left[T_{\gamma\delta}\gamma_{\delta\varepsilon\alpha\beta}^{A}(i\omega,0,0)T_{\varepsilon\eta}\alpha_{\eta\gamma}^{B}(i\omega)\right]. \tag{4.1.7}$$

The contributions $\sim R^{-(>6)}$ to $\alpha_{\alpha\beta}^{disp,AB}$ can be obtained from the next terms of (2.3.5), when the polarizabilities in them, considered as a function of the external electric field F_{α}^{0}, are expanded in a Taylor series on F_{α}^{0}.

Unfortunately, direct calculation of the dispersion contribution $\alpha_{\alpha\beta}^{disp,AB}$ using Eq. (4.1.7) is often difficult due to the absence of the $\gamma_{\delta\varepsilon\alpha\beta}^{A}(i\omega,0,0)$ values as functions of $i\omega$. However, as for the case of the dipole moment, this dispersion contribution may be estimated using a "constant ratio" approximation (see, for example, [44]). As a result, the following estimation may be obtained

$$\alpha_{\alpha\beta}^{disp} = \frac{T_{\gamma\delta}\gamma_{\delta\varepsilon\alpha\beta}^{B}(0,0,0)T_{\varepsilon\eta}C_{\eta\gamma}}{2\alpha^{B}(0)} + \frac{T_{\gamma\delta}\gamma_{\delta\varepsilon\alpha\beta}^{A}(0,0,0)T_{\varepsilon\eta}D_{\eta\gamma}}{2\alpha^{A}(0)} \tag{4.1.8}$$

where

$$C_{\eta\gamma} = \frac{1}{2\pi}\int\limits_{0}^{\infty}\alpha^{B}(i\omega)\alpha_{\eta\gamma}^{A}(i\omega)d\omega,$$

$$D_{\eta\gamma} = \frac{1}{2\pi}\int\limits_{0}^{\infty}\alpha^{A}(i\omega)\alpha_{\eta\gamma}^{B}(i\omega)d\omega$$

and the relation has been used

$$\gamma_{\delta\varepsilon\alpha\beta}^{A,B}(i\omega,0,0) = \frac{\gamma_{\delta\varepsilon\alpha\beta}^{A,B}(0,0,0)}{2\alpha^{A,B}(0)}\alpha^{A,B}(i\omega).$$

Here $\alpha^{A,B}(i\omega)$ is the mean polarizability of the A (or B) molecule at the imaginary frequency $i\omega$.

Since the imaginary frequency-dependent polarizability for molecules as a rule is known (or can be calculated) the further calculation of the $\alpha_{\alpha\beta}^{disp}$ is not difficult. The coefficients $C_{\eta\gamma}$ and $D_{\eta\gamma}$ may be also related to the dispersion constants $C_6^0 \equiv C_6$ and C_6^2 determined experimentally.

4.1.3 Exchange Contributions. Analytical Form

To calculate the exchange contributions to the static polarizability of a pair of interacting atoms a well-known form for tensor components of the electron polarizability may be used:

$$\alpha_{\alpha\beta}^{AB}(R) = 2 \sum_{m \neq n} \frac{\langle n|\mu_\alpha|m\rangle \langle m|\mu_\beta|n\rangle}{E_m - E_n} \qquad (4.1.9)$$

where the dipole matrix elements $\langle n|\mu_\gamma|m\rangle$ and the electronic energy levels E_k are the functions of the intermolecular distance R. Then, the idea proposed in Sect. 3.1. 3 to take into account the exchange interaction contributions to $\langle n|\mu_\gamma|m\rangle$ for small R can be used again. In this case the exchange interaction contribution into z-component of the dipole moment is represented by (3.1.21) where

$$\delta = \frac{1}{\beta_A} + \frac{1}{\beta_B} + \frac{1}{\beta_m} + \frac{3}{2(\beta_A + \beta_B)} - \frac{1}{\beta_A + \beta_B + 2\beta_m} + 1, \qquad (4.1.10)$$

$$\eta = \frac{3}{4}\beta_A + \frac{3}{4}\beta_B + \frac{1}{2}\beta_m. \qquad (4.1.11)$$

The use of the effective excited electron state $\bar{m}$ allows us to exclude the procedure of summation over the excited electron states m of atoms in the expression for polarizability. Then, replacing β_m by some effective value $\bar{\beta}$ for the effective electron state $\bar{m}$, the expression for the exchange contributions to the polarizability of interacting atoms takes the form [59]

$$\alpha_{zz}^{exch}(R) = B_1(\beta_A, \beta_B, \bar{\beta}, R)R^\delta \exp(-\eta R) + B_2(\beta_A, \beta_B, \bar{\beta}, R)R^{2\delta} \exp(-2\eta R),$$
$$(4.1.12)$$

where the fitting parameters B_1 and B_2 are the functions weakly dependent on R and the values of the parameter $\bar{\beta}$ may be estimated taking into account probabilities of radiation transitions of the atoms A and B, and the energy levels structure of these atoms. Note that the first term in Eq. (4.1.12) gives the similar asymptotic behavior for $\alpha_{zz}^{exch}(R)$ at $R \to \infty$ as for the pair of hydrogen atoms [60]. The exchange contributions α_{xx}^{exch} and α_{yy}^{exch} as well equal zero for considered approximation. However, in a stricter approximation these contributions are to be appeared.

The results obtained for interacting atoms may be also used for interacting molecules bearing in mind that β_A, β_B and $\bar{\beta}$ are the molecular parameters and the configurations of interacting molecules have to be taken into account.

4.2 Polarizabilities of van der Waals Complexes

4.2.1 Polarizabilities of Atom-Atomic Complexes

For atomic complexes the general expressions (4.1.5) and (4.1.8) are essentially simplified and take, up to terms $\sim R^{-6}$ inclusively, the well-known forms [1, 44]

(here, as usually, the atoms are positioned on the z-axis of the Cartesian coordinate system)

$$\alpha_{zz}^{ind}(R) = \frac{4\alpha^A \alpha^B}{R^3} + \frac{4\alpha^A \alpha^B (\alpha^A + \alpha^B)}{R^6}, \tag{4.2.1}$$

$$\alpha_{xx}^{ind}(R) = \alpha_{yy}^{ind}(R) = -\frac{2\alpha^A \alpha^B}{R^3} + \frac{\alpha^A \alpha^B (\alpha^A + \alpha^B)}{R^6}, \tag{4.2.2}$$

and

$$\alpha_{zz}^{disp}(R) = \frac{7}{18}\left(\frac{\gamma^A}{\alpha^A} + \frac{\gamma^B}{\alpha^B}\right)\frac{C_6^0}{R^6}, \tag{4.2.3}$$

$$\alpha_{xx}^{disp}(R) = \alpha_{yy}^{disp}(R) = \frac{2}{9}\left(\frac{\gamma^A}{\alpha^A} + \frac{\gamma^B}{\alpha^B}\right)\frac{C_6^0}{R^6}, \tag{4.2.4}$$

where α^A, α^B are the polarizabilities and γ^A, γ^B are the second hyperpolarizabilities of the interacting atoms A and B ($\gamma = \gamma_{zzzz}$).

As shown in the work [59] for the atomic complexes He–He, Ar–Ar, Kr–Xe, and Xe–Xe, taking into account the exchange polarizability $\alpha_{zz}^{exch}(R)$ in the form (3.1.12) allows us to reach a very good agreement with ab initio results [61, 62] for a wide range of R. Herewith, the exchange polarizability $\alpha_{zz}^{exch}(R)$ is mainly determined by the first term in Eq. (4.1.12) for the range $R > R_e$ and by the second term for the range $R < R_e$.

The exchange contribution $\alpha_{zz}^{exch}(R)$ to the interaction mean polarizability

$$\Delta\alpha(R) = \frac{1}{3}(\Delta\alpha_{zz}(R) + 2\Delta\alpha_{xx}(R))$$
$$= \frac{2\alpha^A \alpha^B (\alpha^A + \alpha^B)}{R^6} + \frac{5}{18}\left(\frac{\gamma^A}{\alpha^A} + \frac{\gamma^B}{\alpha^B}\right)\frac{C_6}{R^6} + \frac{1}{3}\alpha_{zz}^{exch}(R) \tag{4.2.5}$$

and the interaction polarizability anisotropy

$$\Delta\gamma(R) = \Delta\alpha_{zz}(R) - \Delta\alpha_{xx}(R)$$
$$= \frac{6\alpha^A \alpha^B}{R^3} + \frac{3\alpha^A \alpha^B (\alpha^A + \alpha^B)}{R^6} + \frac{1}{6}\left(\frac{\gamma^A}{\alpha^A} + \frac{\gamma^B}{\alpha^B}\right)\frac{C_6}{R^6} + \alpha_{zz}^{exch}(R) \tag{4.2.6}$$

also plays an important role, especially for $\Delta\alpha(R)$ because of the term $\sim R^{-3}$ (Fig. 4.1 illustrates this effect).

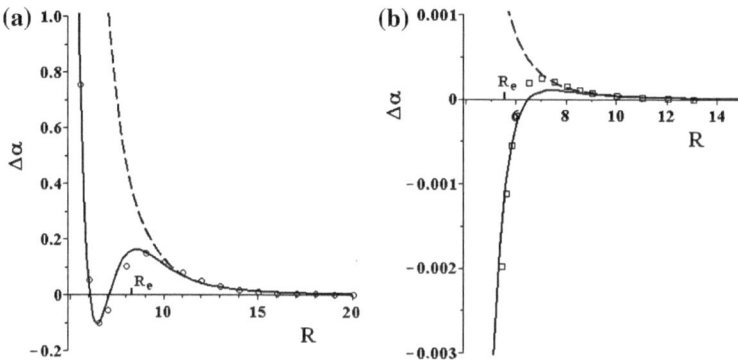

Fig. 4.1 The interaction polarizability invariants $\Delta\alpha(R)$ of the complexes Xe–Xe (**a**) and He–He (**b**) (firstly printed in our work [59]). *Solid lines*—analytical calculations taking into account the exchange polarizability; *dashed lines*—analytical calculations without considering the exchange polarizability; *circles*—ab initio calculation [112]; *squares*—ab initio calculation [62]. All values are in a.u

4.2.2   Polarizabilities of X_2–Y Complexes

Consider in this Section the complexes X_2–Y when the atom Y has the spherical symmetry. For another case the calculations are carried out the same way. The Cartesian coordinate system is introduced following to Fig. 3.1. Then, for considered complex the induction contributions to $\Delta\alpha_{\alpha\beta}^{AB}(R)$ up to terms $\sim R^{-6}$ inclusive take the form [45]

$$\alpha_{\alpha\beta}^{ind} = \alpha_{\alpha\gamma}^A T_{\gamma\delta}\alpha_{\delta\beta}^B + \alpha_{\alpha\gamma}^B T_{\gamma\delta}\alpha_{\delta\beta}^A + \frac{1}{15}\alpha_{\alpha\gamma}^A T_{\gamma\delta\varepsilon\varphi}E_{\beta,\delta\varepsilon}^B + \frac{1}{15}E_{\alpha,\gamma\delta\varepsilon}^B T_{\gamma\delta\varepsilon\varphi}\alpha_{\varphi\beta}^A$$

$$- \frac{1}{9}B_{\alpha\beta,\gamma\delta}^A T_{\gamma\delta\varepsilon\varphi}\Theta_{\varepsilon\varphi}^B + \alpha_{\alpha\gamma}^A T_{\gamma\delta}\alpha_{\delta\varepsilon}^B T_{\varepsilon\varphi}\alpha_{\varphi\beta}^A + \alpha_{\alpha\gamma}^A T_{\gamma\delta}\alpha_{\delta\varepsilon}^A T_{\varepsilon\varphi}\alpha_{\varphi\beta}^B. \quad (4.2.7)$$

Dispersion contribution into $\Delta\alpha_{\alpha\beta}^{AB}(R)$ following the results of Sect. 4.1.2 with accuracy up to the leading term $\sim R^{-6}$ comes on (4.1.8). For this case the tensors $C_{\eta\eta} = \frac{1}{6}C_6^0\delta_{\eta\eta}$, $D_{xx} = D_{yy} = \frac{1}{6}\left(C_6^0 - C_6^2\right)$ and $D_{zz} = \frac{1}{6}\left(C_6^0 + 2C_6^2\right)$ are related to the isotropic dispersion coefficient C_6^0 and the anisotropic dispersion coefficients C_6^2.

The exchange contributions to the interaction polarizabilities are described by Eq. (4.1.12) where the parameters B_1 and B_2 are the functions of the angle θ and can be written as [59]

$$B_i = B_i^{(0)} + B_i^{(2)}P_2(\cos\theta) + B_i^{(4)}P_4(\cos\theta) \quad (4.2.8)$$

where $P_\lambda(\cos\theta)$ is the Legendre polynomial and $B_i^{(\lambda)}$ are the fitting parameters. The full analytical expressions for the interaction polarizability of the complex X–Y_2 is quite cumbersome and can be found in the work [59].

As there was noted in our work [59] the exchange contribution into the components of the complex X–Y_2 can be appreciable for the interaction polarizability $\Delta\alpha_{zz}^{AB}(R,\theta)$. Figure 4.2 illustrates this fact for the case of Ar–H_2 complex.

Modified DID model. As seen from Eq. (4.1.5) for some cases the simple dipole-induced-dipole (DID) model formulated by Silberstein for a system of n interacting atoms [63], being a little modified [50] for the molecular complexes, can be effective applied. In the modified model, like for the dipole moment, every molecule in a complex is represented as a set of effective atoms whose polarizability depends on the internuclear separations in the molecule. The polarizabilities of the effective atoms are chosen so that their total polarizability equals to the polarizability of the molecule formed by these atoms, and the interaction between the effective atoms is absent. The concept of effective atoms allows one to take into account the dependence of the polarizability of a complex on the intramolecular separation r and, because the small interatomic separations are excluded from calculation procedure, the use of usual DID model is appeared.

It is interesting that in the framework of modified DID model for some stable configurations of X–Y_2 complexes the strongest dipole-induced-dipole contributions can be fully summarized. Consider, for example, the complexes N_2–Y and O_2–Y (Y = He, Ne, Ar, Kr, Xe). The calculations of the potential energy surfaces for these complexes show that the complexes N_2–Y exist in one stable configuration (*T*-configuration) [64–77], while the complexes O_2...Y exist in two stable configurations (*L*- and *T*-configurations) [78–82], and the *T*-configuration is the most stable among them. The stable configurations of these complexes are shown in Fig. 4.3, in which atoms 1 and 2 belong to the molecule X_2, and atom 3 is the atom Y.

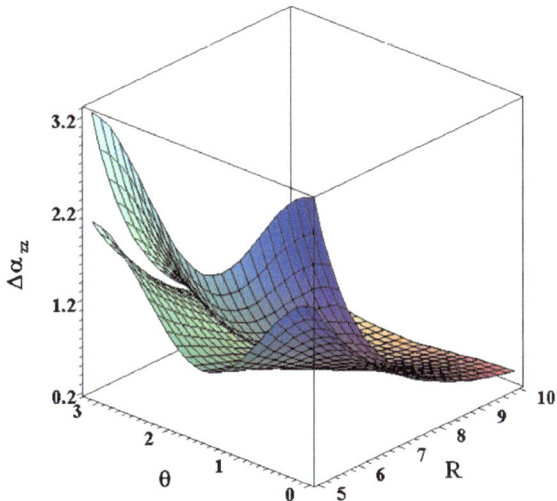

Fig. 4.2 The interaction polarizabilities $\Delta\alpha_{zz}^{AB}(R,\theta)$ of the complex Ar–H_2 with (lower surface) and without (upper surface) taking into account the exchange polarizability (firstly printed in our work [59]). The values of $\Delta\alpha_{zz}^{AB}$ and R are in a.u., the angle θ is in rad

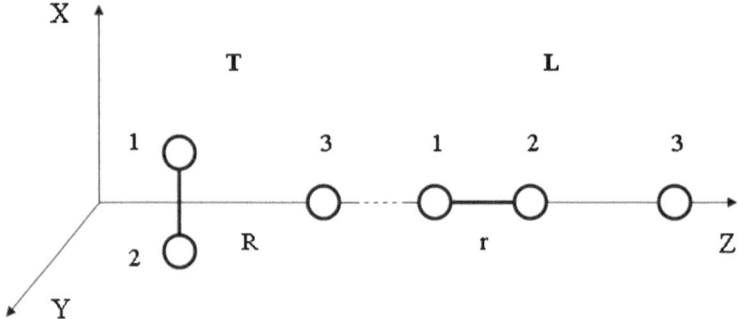

Fig. 4.3 T- and L-configurations of X_2–Y complexes

L-configuration. In this case, the complex-fixed coordinate system (X, Y, Z; see Fig. 4.3) is the main one for the polarizability tensor of the complex, and therefore in this coordinate system it has the diagonal form $\alpha_{\beta\beta}(r, R)$. A feature of the L-configuration of the complex X_2–Y is that the chosen coordinate system coincides with the molecular coordinate system (x, y, z) of the molecule X_2, and for this molecule $\alpha_{XX}(r) = \alpha_{YY}(r) = \alpha_{xx}(r) = \alpha_{yy}(r)$ and $\alpha_{ZZ}(r) = \alpha_{zz}(r)$. And the following analytical explicit forms for the components of the polarizability tensor for the complex X_2–Y in the L-configuration can be found [51]:

$$\alpha_{XX}(r,R) = \alpha_{YY}(r,R) = \frac{\alpha_{xx}(r) + \alpha - \alpha\alpha_{xx}(r)\left(\frac{1}{R_{13}^3} + \frac{1}{R_{23}^3}\right) - \frac{1}{4}\alpha[\alpha_{xx}(r)]^2\left(\frac{1}{R_{13}^3} - \frac{1}{R_{23}^3}\right)^2}{1 - \frac{1}{2}\alpha\alpha_{xx}(r)\left(\frac{1}{R_{13}^6} + \frac{1}{R_{23}^6}\right)},$$

$$(4.2.9)$$

$$\alpha_{ZZ}(r,R) = \frac{\alpha_{zz}(r) + \alpha + 2\alpha\alpha_{zz}(r)\left(\frac{1}{R_{13}^3} + \frac{1}{R_{23}^3}\right) - \alpha[\alpha_{zz}(r)]^2\left(\frac{1}{R_{13}^3} - \frac{1}{R_{23}^3}\right)^2}{1 - 2\alpha\alpha_{zz}(r)\left(\frac{1}{R_{13}^6} + \frac{1}{R_{23}^6}\right)}, \quad (4.2.10)$$

where $R_{13} = R + \frac{r}{2}$, $R_{23} = R - \frac{r}{2}$, α is isotropic polarizability tensor of atom Y. As a result, the expressions (4.2.9) and (4.2.10) allow us to take into account the r-dependence of the polarizability components of the complex.

T-configuration. In this case, the chosen coordinate system (Fig. 4.3) is also the main one for the polarizability tensor of the complex, but it does not coincide with the molecular coordinate system of the molecule X_2. However, for these coordinate systems there is a unique correspondence between the components of the polarizability tensor of the molecule X_2: $\alpha_{XX}(r) = \alpha_{zz}(r)$ и $\alpha_{YY}(r) = \alpha_{ZZ}(r) = \alpha_{xx}(r)$. And the calculation of the components of the polarizability tensor of the complex X_2–Y in the T-configuration yields the following analytical expressions [51]:

$$\alpha_{XX}(r,R) = \frac{\alpha_{zz}(r) + \alpha + 2\alpha\alpha_{zz}(r)\frac{\frac{r^2}{2}-R^2}{\left(R^2+\frac{r^2}{4}\right)^{5/2}} - \frac{9}{4}\alpha\alpha_{xx}(r)\alpha_{zz}(r)\frac{r^2R^2}{\left(R^2+\frac{r^2}{4}\right)^5}}{1 - \alpha\alpha_{zz}(r)\frac{\left(\frac{r^2}{2}-R^2\right)^2}{\left(R^2+\frac{r^2}{4}\right)^5} - \frac{9}{4}\alpha\alpha_{xx}(r)\frac{r^2R^2}{\left(R^2+\frac{r^2}{4}\right)^5}}, \quad (4.2.11)$$

$$\alpha_{YY}(r,R) = \frac{\alpha_{xx}(r) + \alpha - 2\alpha\alpha_{xx}(r)\frac{1}{\left(R^2+\frac{r^2}{4}\right)^{3/2}}}{1 - \alpha\alpha_{xx}(r)\frac{1}{\left(R^2+\frac{r^2}{4}\right)^3}}, \quad (4.2.12)$$

$$\alpha_{ZZ}(r,R) = \frac{\alpha_{xx}^M(r) + \alpha + 2\alpha\alpha_{xx}(r)\frac{2R^2-\frac{r^2}{4}}{\left(R^2+\frac{r^2}{4}\right)^{5/2}} - \frac{9}{4}\alpha\alpha_{xx}(r)\alpha_{zz}(r)\frac{r^2R^2}{\left(R^2+\frac{r^2}{4}\right)^5}}{1 - \alpha\alpha_{xx}(r)\frac{\left(2R^2-\frac{r^2}{4}\right)^2}{\left(R^2+\frac{r^2}{4}\right)^5} - \frac{9}{4}\alpha\alpha_{zz}(r)\frac{r^2R^2}{\left(R^2+\frac{r^2}{4}\right)^5}}. \quad (4.2.13)$$

Thus, the polarizability of the stable configurations of the complex X_2–Y is determined by the polarizability tensor $\alpha_{\alpha\alpha}(r)$ of the molecule X_2, dependent on its internuclear separation r, the polarizability of the noble gas atom α, and the distance R. Note that at $r = 0$ Eqs. (4.2.9)–(4.2.13) coincide with the corresponding equations for the components of the polarizability tensor of two interacting anisotropic atoms in the classical Silberstein theory. Figure 4.4 shows the typical forms of surfaces for the components of the polarizability of the complexes N_2–Y and O_2–Y (Y = He, Ne, Ar, Kr, Xe) calculated in the framework this model by the use of the polarizability functions $\alpha_{xx}(r)$ and $\alpha_{zz}(r)$ for the molecules N_2 and O_2 taken from [83]. Figures 4.5 and 4.6 show also the frequency dependences of the polarizabilities for these complexes. In order to calculate the dynamic polarizabilities of the complexes, the dynamic polarizability functions $\alpha_{xx}(r_e,\omega) = \alpha_{xx}(r_e,\omega)$ and $\alpha_{zz}(r_e,\omega)$ of the molecules N_2 and O_2 are taken from [84, 85] and the dynamic polarizabilities of noble gas atoms $\alpha(\omega)$ are taken from [86–88].

The dependence of the polarizability tensor invariants of the van der Waals X_2–Y complexes on the frequency of the electromagnetic field and on the complex

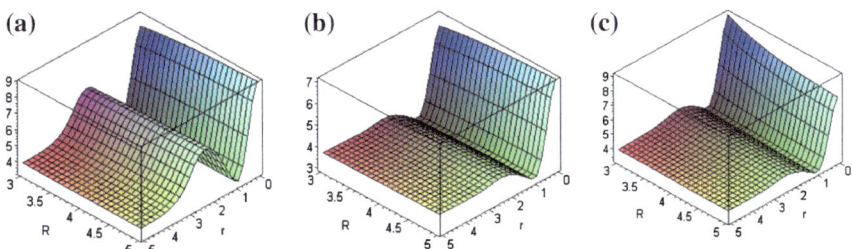

Fig. 4.4 Components $\alpha_{\alpha\alpha}(r,R)$ of the polarizability tensor for the N_2–Ar complex in the T-configuration (in Å^3) [51]. Here the distances r and R are given in Å: **a**—$\alpha_{XX}(r,R)$, **b**—$\alpha_{YY}(r,R)$, **c**—$\alpha_{ZZ}(r,R)$

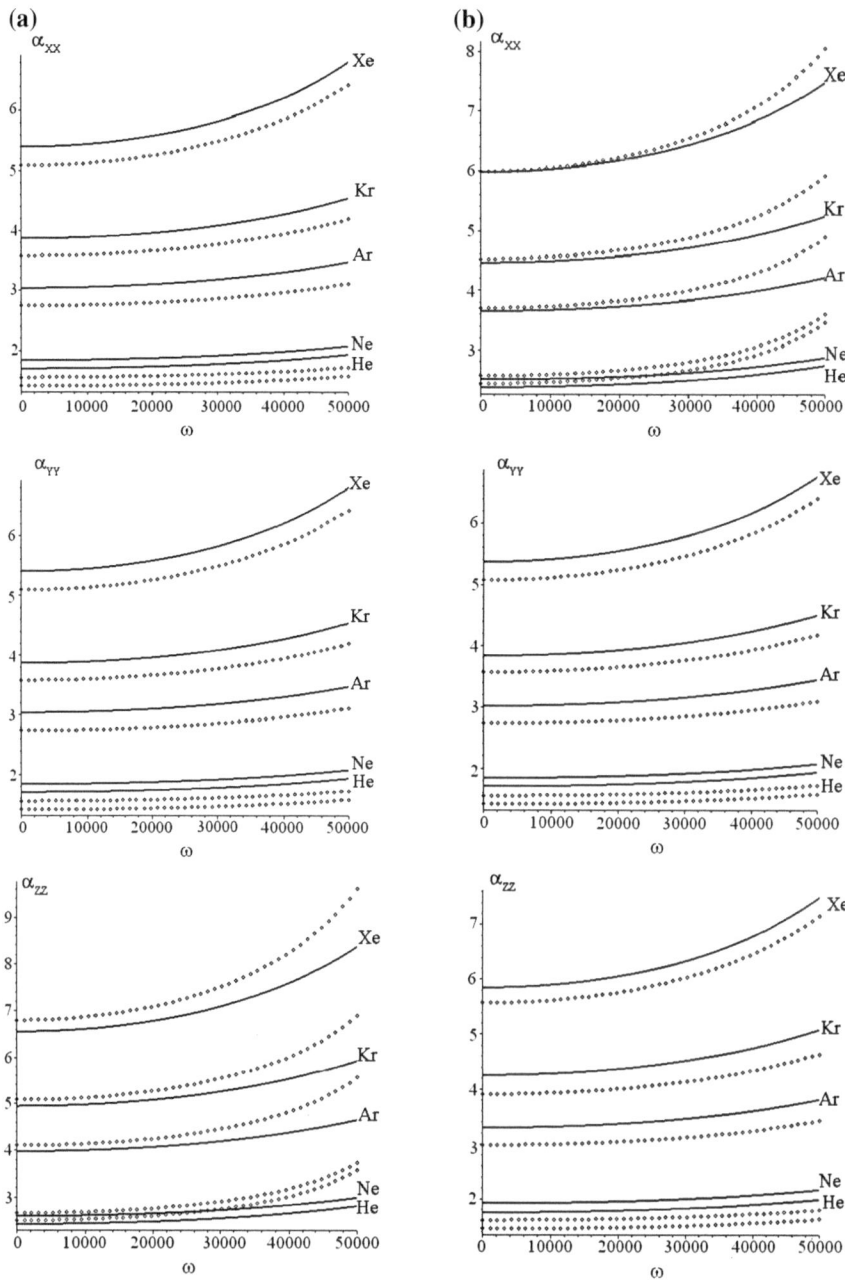

Fig. 4.5 Components of the dynamic polarizability tensor (in Å^3) of the complexes N$_2$–Y (*solid curves*) and O$_2$–Y (*dashed curves*) (Y = He, Ne, Ar, Kr, Xe) [53]: **a** *L*-configuration ($\theta = 0°$); **b** *T*-configuration ($\theta = 90°$); the frequency ω is given in cm^{-1}

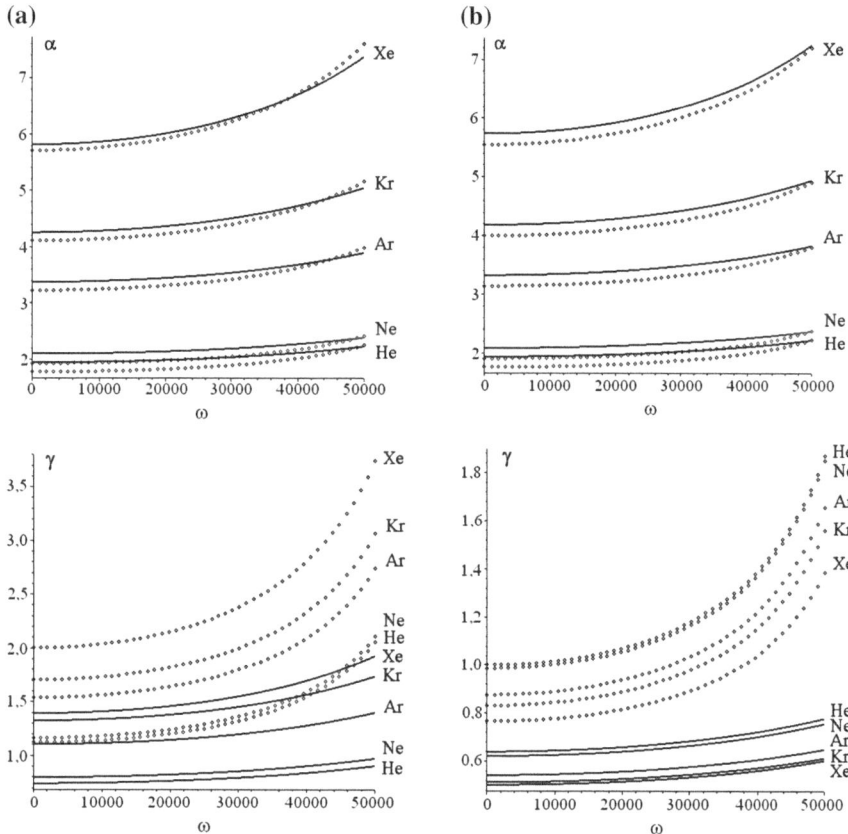

Fig. 4.6 Invariants of the dynamic polarizability tensor $\alpha(re, Re, \theta, \omega)$ and $\gamma(re, Re, \theta, \omega)$ (in Å^3) of the complexes N_2–Y (*solid curves*) and O_2–Y (*dashed curves*) (Y = He, Ne, Ar, Kr, Xe) [53]: **a** *L*-configuration ($\theta = 0°$), **b** *T*-configuration ($\theta = 90°$); the frequency ω is given in cm^{-1}

configuration can be represented more clearly by the polarizability surfaces. As an example, Fig. 4.7 shows such surfaces for the complex O_2–Ar. The polarizability surfaces for other complexes under consideration have analogous shape. It is obvious, that at $\theta = 0$ and $90°$ these surfaces are degenerated into the dynamic polarizability functions shown in Fig. 4.6.

4.2.3   Polarizabilities of X_2–Y_2 Complexes

The polarizabilities of two interacting diatomic molecules are also of interest up to day, however, there are only few works on this subject. Mainly, these works were devoted to some atmospheric complexes like O_2–O_2, N_2–N_2, H_2–H_2, H_2–N_2

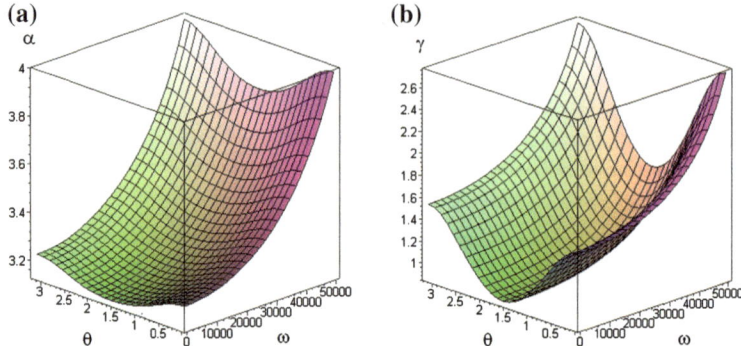

Fig. 4.7 Invariants of the dynamic polarizability tensor $\alpha(r_e, R_e, \theta, \omega)$ and $\gamma(r_e, R_e, \theta, \omega)$ (in Å^3) of the complex O_2–Ar [53]; the angle θ is given in rad; the frequency ω is given in cm^{-1}

systems [48, 50, 54, 89–94] or NO dimer [95]. Consider below the less-studied problem of influence of nuclear vibrations on the polarizability of such dimers.

A. *Influence of nuclear vibrations on the $(N_2)_2$ and $(O_2)_2$ dimer polarizability*

We are restricted here by the terms in (4.2.7) related to the modified DID model discussed above. In the framework of this model for $(N_2)_2$ and $(O_2)_2$ complexes the effective polarizability of the atoms N and O is determined as a half of the polarizability for the N_2 and O_2 molecules respectively. It's evident, that the effective polarizability of atoms N and O determined by such a way is anisotropic and depends on the internuclear distance of the molecules, and, as a result, the polarizability of the complex is the function of the internuclear distances of the molecules forming it, which allows one to calculate the tensor of the polarizability derivatives of the complex for different configurations.

The polarizability tensor of these free oriented interacting molecules depends on the Euler angles $\theta_1, \theta_2, \varphi = \varphi_1 - \varphi_2$ for both molecules, the intermolecular distance R, and the internuclear distances in the molecules r_1 and r_2. For considered model the total induced dipole moment of atomic system is written as

$$\mu_\alpha = \sum_m \mu_\alpha^m = \alpha_{\alpha\beta} E_\beta^0, \tag{4.2.14}$$

where μ_α^m is the induced dipole moment of the atom m and the polarizability of the complex included N atoms is

$$\alpha_{\alpha\beta} = \sum_{m=1}^{N} \alpha_{\alpha\beta}^m + \sum_{m,n=1}^{N} \alpha_{\alpha\delta}^m T_{\delta\gamma}^{mn} \alpha_{\gamma\beta}^n + \sum_{m,n,k=1}^{N} \alpha_{\alpha\delta}^m T_{\delta\gamma}^{mn} \alpha_{\gamma\varepsilon}^n T_{\varepsilon\rho}^{nk} \alpha_{\rho\beta}^k + \cdots. \tag{4.2.15}$$

The Euler angles determine the orientation of the first and the second molecules relative to the coordinate system related to the molecular complex. In the case of diatomic molecules each component of the polarizability tensor of two interacting diatomic molecules is a surface in the space of the variables $\theta_1, \theta_2, \varphi, r_1, r_2$ (Z-axis is directed through the centers of mass of interacting molecules). The analytical description of this surface is cumbersome in the general case. For the specific configurations of the interacting molecules the components of the polarizability tensor depend on R only and may be represented as

$$\alpha_{ij}(r_1, r_2, R) = c_{ij}^{(0)}(r_1, r_2) + \frac{c_{ij}^{(3)}(r_1, r_2)}{R^3} + \frac{c_{ij}^{(5)}(r_1, r_2)}{R^5} + \frac{c_{ij}^{(6)}(r_1, r_2)}{R^6}$$
$$+ \frac{c_{ij}^{(7)}(r_1, r_2)}{R^7} + \frac{c_{ij}^{(8)}(r_1, r_2)}{R^8} + \cdots, \qquad (4.2.16)$$

where the coefficients $c_{ij}^{(k)}(r_1, r_2)$ depend on the relative orientation of the interacting molecules and their intramolecular distances r_1 and r_2. Expanding the function $\alpha_{ij}(r_1, r_2, R)$ into the Taylor series at the equilibrium points r_1^0 and r_2^0 of the first and the second molecules, the following expression for the tensor of the first derivatives of polarizability $\alpha'_{ij}(R) = \left[\partial \alpha_{ij}(r_1, r_2, R)/\partial \xi\right]_{r_1=r_1^0, r_2=r_2^0}$ for the complex may be written as

$$\alpha'_{ij}(R) = d_{ij}^{(0)} + \frac{d_{ij}^{(3)}}{R^3} + \frac{d_{ij}^{(5)}}{R^5} + \frac{d_{ij}^{(6)}}{R^6} + \frac{d_{ij}^{(7)}}{R^7} + \frac{d_{ij}^{(8)}}{R^8} + \cdots, \qquad (4.2.17)$$

where $d_{ij}^{(k)} = \left[\partial c_{ij}^{(k)}(r_1, r_2)/\partial \xi\right]_{r_1=r_1^0, r_2=r_2^0}$ and $\xi = (r - r^0)/r^0$.

The results of calculations of the tensor components for the first derivatives of the polarizability for the N_2 and O_2 molecules, being on the left side in configurations H, X, T, T* and L of considered dimers (Fig. 4.8), are given in Fig. 4.9.

Fig. 4.8 Some configurations of dimers for diatomic molecules

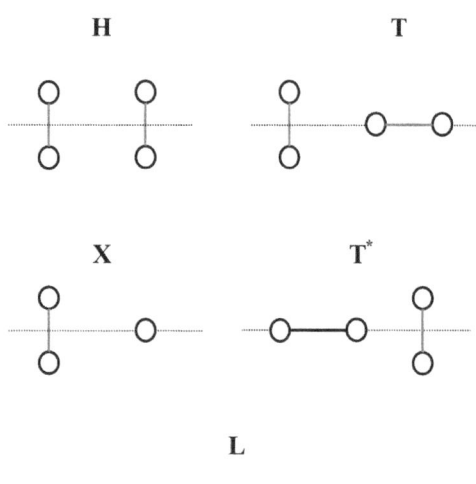

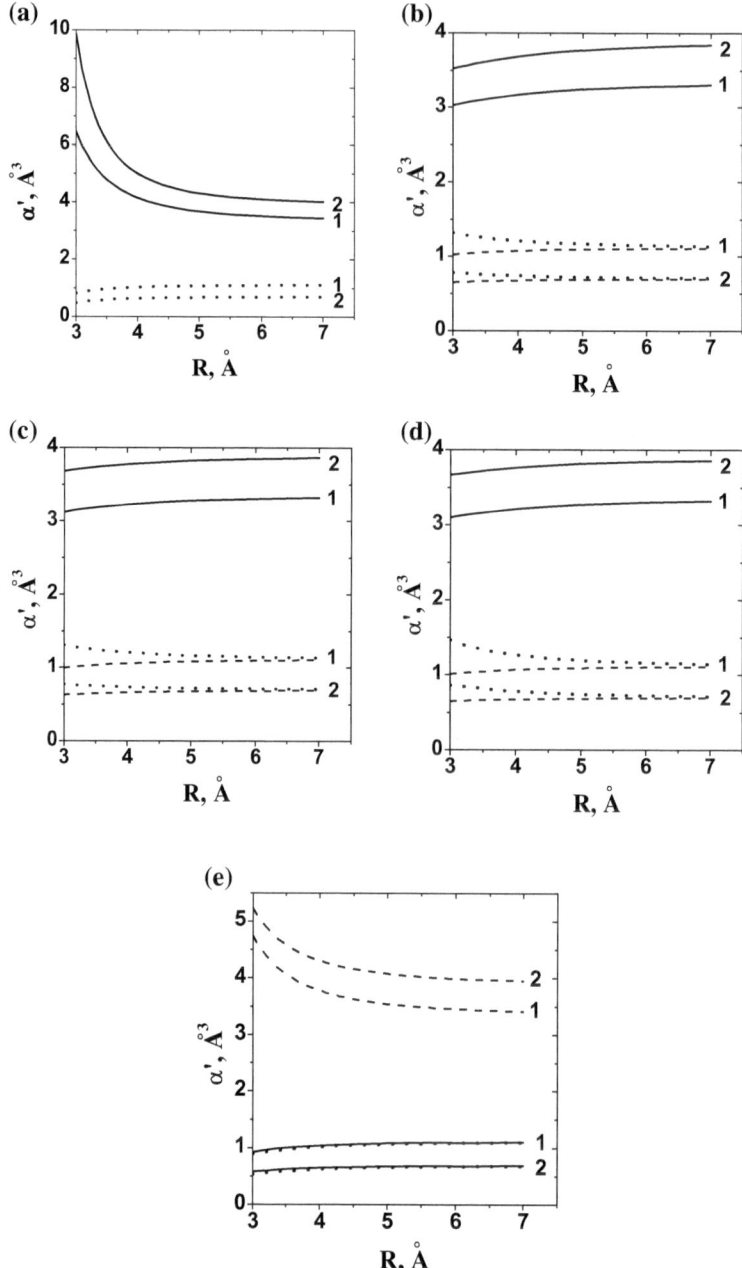

Fig. 4.9 The first derivative of the polarizability for the molecules N_2 (1) and O_2 (2) in dimers: **a** —L configuration, *solid lines*—α'_{zz}, *dotted lines*—$\alpha'_{xx} = \alpha'_{yy}$; **b**—H configuration, *solid lines*—α'_{zz}, *dotted lines*—α'_{xx}, *dashed lines*—α'_{yy}; **c**—X configuration, *solid lines*—α'_{zz}, *dotted lines*—α'_{xx}, *dashed lines*—α'_{yy}; **d**—T configuration, *solid lines*—α'_{zz}, *dotted lines*—α'_{xx}, *dashed lines*—α'_{yy}; **e**— T^* configuration, *dashed lines*—α'_{zz}, *dotted lines*—α'_{xx}, *solid lines*—α'_{yy}

The tensor components of the effective polarizability and its first derivatives for the atoms N and O are the half of the respective tensor components of the polarizability tensors and their first derivatives [96] for the molecules N_2 and O_2: $\alpha_{xx}^{(N)} = \alpha_{yy}^{(N)} = 0.767$ Å^3, $\alpha_{zz}^{(N)} = 1.122$ Å^3, $\alpha_{xx}^{(O)} = \alpha_{yy}^{(O)} = 0.625$ Å^3, $\alpha_{zz}^{(O)} = 1.165$ Å^3 and $\alpha_{xx}'^{(N)} = \alpha_{yy}'^{(N)} = 0.558$ Å^3, $\alpha_{zz}'^{(N)} = 1.673$ Å^3, $\alpha_{xx}'^{(O)} = \alpha_{yy}'^{(O)} = 0.348$ Å^3, $\alpha_{zz}'^{(O)} = 1.943$ Å^3 (the small subscripts x, y, z are the Cartesian coordinates related to the molecule N_2 or O_2 respectively, the axis z is directed along the molecule axis). As it's seen from the Fig. 4.9 the tensor components α_{xx}' and α_{yy}' differ from each other with decreasing of the intermolecular distance (except the configuration L). Herewith, the difference between the components α_{xx}' and α_{yy}' is increased with decreasing of R. The components α_{zz}' have strongest changes for the configurations L and T*, moreover, their values are increasing with decreasing of R.

Note that the relative changes of tensor components for the first derivatives of the polarizability $\delta\alpha_{ii}'(R_e) = \left[\alpha_{ii}'(R_e) - \alpha_{ii}'(R = \infty)\right] / \alpha_{ii}'(R = \infty)$ for the molecules N_2 and O_2 in the dimers $(N_2)_2$ and $(O_2)_2$, where $\alpha_{ii}'(R = \infty)$ are the values for the free molecule, are great enough and vary in the range of 3–15 %.

The approach considered here to calculate the tensor components for the first derivatives of the polarizability for the interacting diatomic molecules can be applied to larger molecular complexes. Also, it may be useful to calculate the tensor components of higher derivatives of the polarizability. The values of the tensor components of the first derivatives of the polarizability for molecules N_2 and O_2 may be used to estimate the parameters of the light Raman scattering in the dimers.

4.2.4 Polarizability of CH₄–N₂ Complex

As mentioned above in Sect. 3.2.3 the molecular complex CH_4–N_2 is of particular interest for planetary applications. However, there is now very scanty information about the polarizability surface of CH_4–N_2 complex especially for a wide range of intermolecular separations and mutual orientations of the interacting molecules [97].

A. *Polarizability surface*

In the work [97] (see the parameters of configurations and the molecular parameters used for calculations in Sects. 3.2.3 and 5.3: Fig. 3.6, Table 5.2) ab initio calculations for the CH_4–N_2 complex were carried out by means of the finite-field method based on the finite differences [98] using Gaussian 03 package [113] at the MP2 and CCSD(T) levels of theory with a correlation-consistent aug-cc-pVTZ basis set and taking into account the BSSE correction. We have to note that the use of the MP2 level of theory gives a good description of the polarizability at low computational costs compared to the CCSD(T) level of theory. Indeed, the polarizability components, for example, for the most stable configuration of the CH_4–N_2 complex calculated at the MP2 and CCSD(T) levels of theory with the aug-cc-pVTZ basis set differ of less than ~ 1 %.

Results of ab initio calculations for the polarizability tensor components of the CH_4–N_2 complex are shown at Fig. 4.10. As expected, the dependence of the polarizability tensor components of this complex on intermolecular distance R has the usual form. However, it is also seen that the polarizability of the complex is affected by the orientation of the N_2 molecule due to its big anisotropy of the polarizability. The orientation of highly symmetric molecule CH_4 in the complex affects significantly smaller on the complex polarizability and becomes noticeable only at small intermolecular separations. It should be pointed out that for configurations 4 and 5 (symmetry C_s) the small nondiagonal component α_{xy} ($\alpha_{xy} \sim 0.005$ a.u. at $R = 3.5$ Å) is appeared in the coordinate system used (Fig. 3.6a).

The analytical calculations in the framework of the long-range approximation can be carried out following the Sect. 4.1.2. Then, for the CH_4–N_2 complex, using the symmetry properties of the molecules CH_4 (A) and N_2 (B), the induction term in Eq. (4.2.15) takes the form

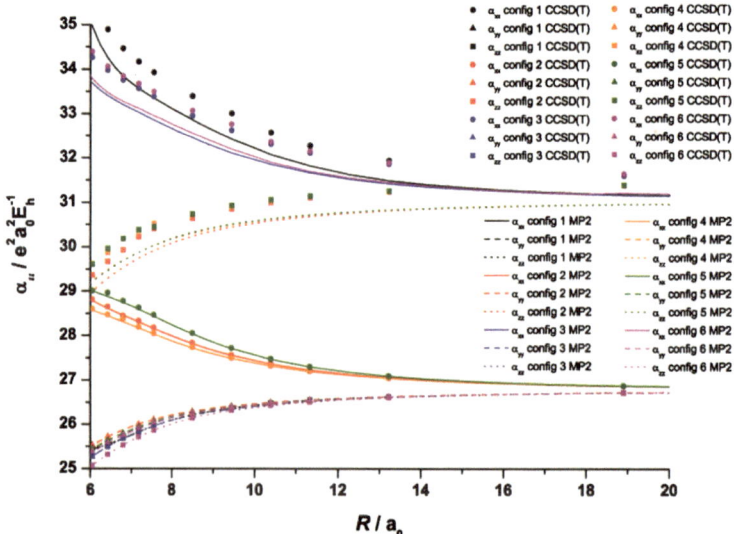

Fig. 4.10 Polarizability components α_{ii} of the CH_4–N_2 complex calculated at the CCSD(T) and MP2 levels of theory with aug-cc-pVTZ basis set with the BSSE correction [97]. All values are in a.u. *Black color*—conf. 1, *red color*—conf. 2, *blue color*—conf. 3, *orange color*—conf. 4, *olive color*—conf. 5, *magenta color*—conf. 6; *solid lines*—α_{xx} (MP2), *dash lines*—α_{yy} (MP2), *dot lines*—α_{zz} (MP2); *circles*—α_{xx} (CCSD(T)), *triangles*—α_{yy} (CCSD(T)), *squares*—α_{zz} (CCSD(T)). (Reprinted with permission from Ref. [97]. Copyright 2010 American Institute of Physics.)

$$\alpha_{\alpha\beta}^{ind} = \alpha_{\alpha\gamma}^A T_{\gamma\delta}\alpha_{\delta\beta}^B + \alpha_{\alpha\gamma}^B T_{\gamma\delta}\alpha_{\delta\beta}^A + \frac{1}{3}\beta_{\alpha\beta\gamma}^A T_{\gamma\delta}\Theta_{\delta\varepsilon}^B - \frac{1}{3}A_{\alpha,\gamma\delta}^A T_{\gamma\delta\varepsilon}\alpha_{\varepsilon\beta}^B - \frac{1}{3}\alpha_{\alpha\gamma}^B T_{\gamma\delta\varepsilon}A_{\beta,\delta\varepsilon}^A$$

$$+ \frac{1}{15}\alpha_{\alpha\gamma}^A T_{\gamma\delta\varepsilon\varphi}E_{\beta,\delta\varepsilon\varphi}^B + \frac{1}{15}E_{\alpha,\gamma\delta\varepsilon}^B T_{\gamma\delta\varepsilon\varphi}\alpha_{\varphi\beta}^A + \frac{1}{15}E_{\alpha,\gamma\delta\varepsilon}^A T_{\gamma\delta\varepsilon\varphi}\alpha_{\varphi\beta}^B + \frac{1}{15}\alpha_{\alpha\gamma}^B T_{\gamma\delta\varepsilon\varphi}E_{\beta,\delta\varepsilon\varphi}^A$$

$$- \frac{1}{9}B_{\alpha\beta,\gamma\delta}^A T_{\gamma\delta\varepsilon\varphi}\Theta_{\varepsilon\varphi}^B + \alpha_{\alpha\gamma}^A T_{\gamma\delta}\alpha_{\delta\varepsilon}^B T_{\varepsilon\varphi}\alpha_{\varphi\beta}^A + \alpha_{\alpha\gamma}^B T_{\gamma\delta}\alpha_{\delta\varepsilon}^A T_{\varepsilon\varphi}\alpha_{\varphi\beta}^B + \cdots \qquad (4.2.18)$$

In Eq. (4.2.18) the long-range multipolar induction contributions up to terms $\sim R^{-5}$ inclusively and terms $\sim R^{-6}$ caused by back induction effect due to dipole-induced-dipole interaction have been fully accounted. The contribution of dispersion interactions to the polarizability of the interacting molecules can also be calculated in the manner described in Sect. 4.1.2.

For comparison of the results of ab initio and analytical calculations, it is convenient to use the interaction polarizabilities $\Delta\alpha_{\alpha\beta}$ defined as

$$\Delta\alpha_{\alpha\beta} = \alpha_{\alpha\beta}^{AB} - \alpha_{\alpha\beta}^A - \alpha_{\alpha\beta}^B \equiv \alpha_{\alpha\beta}^{AB}(R) - \alpha_{\alpha\beta}^{AB}(\infty). \qquad (4.2.19)$$

The calculation results of $\Delta\alpha_{\alpha\beta}$ obtained using both methods for two typical configurations 3 and 4 of the complex CH_4-N_2 are given in Fig. 4.11 (Fig. 4.11a: $\alpha_{xx}^{AB} \neq \alpha_{yy}^{AB} = \alpha_{zz}^{AB}$ and Fig. 4.11b: $\alpha_{xx}^{AB} \neq \alpha_{yy}^{AB} \neq \alpha_{zz}^{AB}$). Note that the exchange contribution to the analytical form (4.1.12) in the calculations was not taken into account. It is seen that the values of $\Delta\alpha_{yy}$ and $\Delta\alpha_{zz}$ are in a good agreement for all range of considered distances R, while the values of $\Delta\alpha_{xx}$ agree well only for $R > 10$ a.u. The same result is also obtained for other configurations of the complex CH_4-N_2.

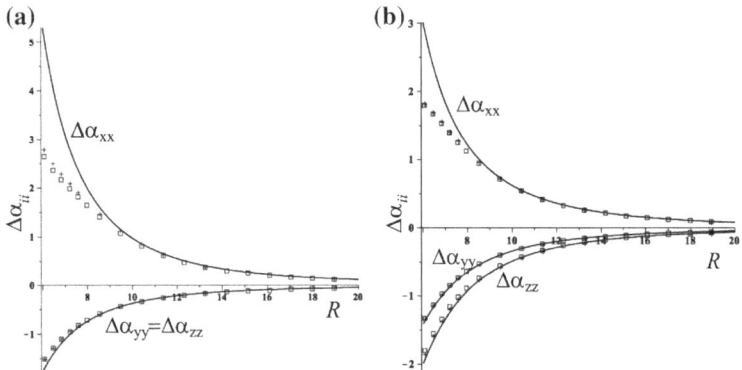

Fig. 4.11 Interaction polarizabilities $\Delta\alpha_{ii}$ of CH_4-N_2 complex (**a**—the configuration 3, **b**—the configuration 4) [97]. All units are in a.u. *Solid lines*—analytical calculations, *crosses*—CCSD(T) calculations, and *boxes*—MP2 calculations. (Reprinted with permission from Ref. [97]. Copyright 2010 American Institute of Physics.)

Analytical calculations allow us also to estimate easily the contributions of the different induction and dispersion interactions to $\Delta\alpha_{ii}$ of the complex CH_4–N_2. Their analysis shows (see details in [97]) that the leading contributions to the polarizability for any complex configuration from Fig. 3.6 are due to dipole-induced-dipole interactions or to terms with $\alpha^A\alpha^B \sim R^{-3}$ in (4.2.18). The other induction terms ($\sim R^{-4}$ ($A^A\alpha^B$) and $\sim R^{-5}$ ($E^A\alpha^B$, $\alpha^A E^B$ and $B^A\theta^B$), and also the terms $\sim R^{-6}$ caused by the dipole-induced-dipole interaction ($\alpha^A\alpha^B\alpha^A + \alpha^B\alpha^A\alpha^B$)) and dispersion terms ($\sim R^{-6}$) are significantly smaller and are of comparable values. It should be noticed that the induction term with $\beta^A\theta^B$ ($\sim R^{-4}$) is very small for all considered configurations.

The calculations of the mean ($\Delta\alpha$) and the anisotropy ($\Delta\gamma$) of the interaction polarizability

$$\Delta\alpha = \alpha^{AB} - \alpha^{CH_4} - \alpha^{N_2} \qquad (4.2.20)$$

and

$$\Delta\gamma = \gamma^{AB} - \gamma^{N_2}, \qquad (4.2.21)$$

which describe the contribution of intermolecular interactions to the mean polarizability α^{AB} and the anisotropy of the polarizability tensor γ^{AB}, are presented in Figs. 4.12 and 4.13. Here, α^{CH_4}, α^{N_2} and γ^{N_2} are the mean polarizabilities and the anisotropy of the molecules CH_4 and N_2. It can be seen that for $R \geq 10\ a_0$ there is a very good agreement between analytical and ab initio calculations. Herewith, the

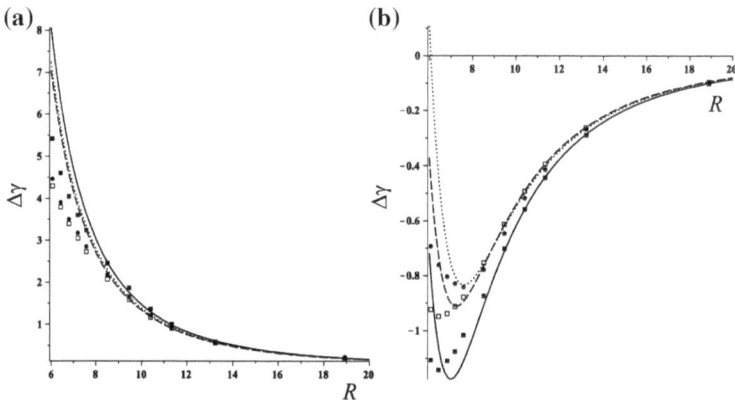

Fig. 4.12 Interaction anisotropy $\Delta\gamma$ for CH_4–N_2 complex (**a**—configurations 1, 3 and 6; **b**—configurations 2, 4 and 5). All values are in a.u. [97]. (Reprinted with permission from Ref. [97]. Copyright 2010 American Institute of Physics.) The configurations 1 and 2: *solid lines*—analytical calculations, *solid boxes*—CCSD (T) calculations. The configurations 3 and 4: *dash lines*—analytical calculations, *boxes*—CCSD (T) calculations. The configurations 5 and 6: *dot lines*—analytical calculations, *solid circles*—CCSD (T) calculations

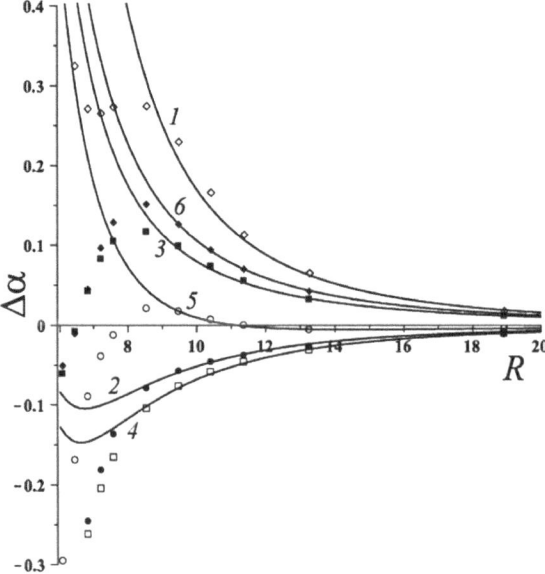

Fig. 4.13 The mean interaction polarizability $\Delta\alpha$ for the CH_4–N_2 complex for configurations 1–6 (Reprinted with permission from Ref. [97]. Copyright 2010 American Institute of Physics.). All values are in a. u. The numbers in the figure correspond to those of the configurations. *Solid lines*— analytical calculations, *diamonds*—CCSD (T) calculations for the configuration 1, *solid circles*— CCSD(T) calculations for the configuration 2, *solid boxes*—CCSD(T) calculations for the configuration 3, *boxes*—CCSD(T) calculations for the configuration 4, *circles*—CCSD(T) calculations for the configuration 5, *solid diamonds*—CCSD(T) calculations for the configuration 6

induction terms $\alpha^A\alpha^B$ $(\sim R^{-3})$ gives the leading contribution to the interaction anisotropy of the complex for $R \geq 10\ a_0$ while for $R \leq 10\ a_0$ the substantial contribution to $\Delta\gamma$ is caused by dispersion and induction terms of higher orders. As to the mean interaction polarizability $\Delta\alpha$, the analytical calculations do not allow to describe correctly the function $\Delta\alpha(R)$ for $R < 10\ a_0$ even qualitatively.

B. *Polarizability of the most stable configuration*

As it was mentioned in Sect. 3.2.2 (*B*) the complex CH_4–N_2 has a family of the most stable configurations. It is obvious that the polarizability tensor components $\alpha_{\alpha\beta}^{AB}(R_e)$ are different for configurations of the family. However, the ab initio calculations of the polarizability tensor invariants $\alpha^{AB}(R_e)$ and $\gamma^{AB}(R_e)$ for $R_e = 6.84\ a_0$ for this family of configurations have shown that the values of the invariants are practically equal to $\alpha^{AB}(R_e) = 28.13$ a.u. and $\gamma^{AB}(R_e) = 3.82$ a.u. These values are less than the values of the invariants $\alpha^{AB}(\infty) = 28.34$ a.u. and $\gamma^{AB}(\infty) = 4.61$ a.u. for the non-interacting molecules CH_4 and N_2. Such reduction of polarizability invariants under formation of the more stable configurations leads to the decrease of the efficiency of light scattering (index of the refraction $n \sim \alpha$) including rotational Raman scattering. Note, that the calculated values $\alpha^{AB}(\infty)$ and $\gamma^{AB}(\infty)$ are in a good agreement with the values $\alpha^{CH_4} + \alpha^{N_2} = 28.16$ a.u. and $\gamma^{N_2} = 4.61$ a.u. obtained by G. Maroulis [99, 100].

It should be noted that at the formation of the complex the monomers CH_4 and N_2 are deformed. As a result, the complex polarizability is changed due to the nonrigidity of the monomers. However, these changes are not large. Indeed, the calculations for the most stable configurations of the CH_4–N_2 complex at full optimization of their geometries have shown that accounting for the monomers nonrigidity practically does not affect the mean polarizability $\alpha^{AB}(R_e)$ and changes slightly (several percent) the polarizability anisotropy $\gamma^{AB}(R_e)$.

Polarizability tensor invariants $\alpha^{AB}(R)$ and $\gamma^{AB}(R)$ for the family of the most stable configurations of the CH_4–N_2 complex can be also presented in the form of Taylor series in the vicinity of $R_e = 6.84\ a_0$:

$$\alpha^{AB}(R) = 27.940 + 0.172(R - 6.84) - 0.075(R - 6.84)^2 + 0.015(R - 6.84)^3,$$
$$(4.2.22)$$

$$\gamma^{AB}(R) = 3.421 - 0.018(R - 6.84) + 0.091(R - 6.84)^2 - 0.020(R - 6.84)^3.$$
$$(4.2.23)$$

These expressions allow estimating the derivatives of the polarizability tensor invariants of the complex for the analysis of scattering processes in methane-nitrogen gas media. Particularly, it is seen that the first derivatives $\partial \alpha^{AB}(R)/\partial R = 0.172$ a.u. and $\partial \gamma^{AB}(R)/\partial R = -0.018$ a.u. are of different sign at R_e and considerably differ in their absolute value.

C. *Polarizability of free orientated interacting molecules*

The polarizability of the free oriented interacting molecules CH_4 and N_2 can be easily calculated if the analytical expressions for polarizabilities $\alpha_{\alpha\beta}^{AB}$ are averaged over Euler angles χ_A, θ_A, ϕ_A, θ_B and ϕ_B. In this case, the polarizability tensor has only two independent components:

$$\bar{\alpha}_{xx}^{AB} = \alpha^A + \alpha^B + \frac{4}{R^3}\alpha^A \alpha^B + \frac{4}{R^6}\alpha^A \alpha^B (\alpha^A + \alpha^B) + \frac{22\alpha^A}{45R^6}(\alpha_{zz}^B - \alpha_{xx}^B)^2 + \frac{7C_6^{AB}}{18R^6}\left(\frac{\bar{\gamma}^A}{\alpha^A} + \frac{\bar{\gamma}^B}{\alpha^B}\right)$$
$$(4.2.24)$$

$$\bar{\alpha}_{yy}^{AB} = \bar{\alpha}_{zz}^{AB} = \alpha^A + \alpha^B - \frac{2}{R^3}\alpha^A \alpha^B + \frac{1}{R^6}\alpha^A \alpha^B (\alpha^A + \alpha^B)$$
$$+ \frac{19\alpha^A}{45R^6}(\alpha_{zz}^B - \alpha_{xx}^B)^2 + \frac{2C_6^{AB}}{9R^6}\left(\frac{\bar{\gamma}^A}{\alpha^A} + \frac{\bar{\gamma}^B}{\alpha^B}\right),$$
$$(4.2.25)$$

where $\bar{\gamma}^A$ and $\bar{\gamma}^B$ are the mean hyperpolarizabilities of the methane and nitrogen molecules that have the form

$$\bar{\gamma}^A = \frac{3}{5}\left(\gamma^A_{zzzz} + 2\gamma^A_{zxzx}\right), \tag{4.2.26}$$

$$\bar{\gamma}^B = \frac{1}{15}\left(3\gamma^B_{zzzz} + 12\gamma^B_{zxzx} + 8\gamma^B_{xxxx}\right). \tag{4.2.27}$$

Note that when $\alpha^B_{xx} = \alpha^B_{zz}$, Eqs. (4.2.24) and (4.2.25) become the well-known expressions for the polarizabilities of two interacting spherically symmetric atoms A and B [43, 44].

It is also seen from (4.2.24) and (4.2.25) that the dependence of $\bar{\alpha}^{AB}_{ii}$ and the anisotropy of the polarizability tensor $\bar{\alpha}^{AB}_{xx} - \bar{\alpha}^{AB}_{yy}$ on R is defined mainly by the terms $\sim R^{-3}$, while the leading terms of the mean polarizability $\bar{\alpha}^{AB} = \left(\bar{\alpha}^{AB}_{xx} + 2\bar{\alpha}^{AB}_{yy}\right)/3$ are proportional to R^{-6}:

$$\bar{\alpha}^{AB} - \alpha^A - \alpha^B = \frac{2}{R^6}\alpha^A\alpha^B\left(\alpha^A + \alpha^B\right) + \frac{4\alpha^A}{9R^6}\left(\alpha^B_{zz} - \alpha^B_{xx}\right)^2 + \frac{5C^{AB}_6}{18R^6}\left(\frac{\bar{\gamma}^A}{\alpha^A} + \frac{\bar{\gamma}^B}{\alpha^B}\right). \tag{4.2.28}$$

The results of calculations carried out in work [97] have also shown that the induction and dispersion terms ($\sim R^{-6}$) of Eq. (4.2.28) give the leading contribution ($\sim 99\%$). The last part of contribution caused by the anisotropy of the polarizability tensor of the N_2 molecule is very small ($\sim 1\%$).

D. *Vibrational Raman spectrum*

Some features of vibrational Raman spectrum for the most stable configuration 4 (see Fig. 3.6) of the CH_4–N_2 complex are discussed in this section following to [101]. This spectrum was obtained using the Gaussian 03 package at the MP2/aug-cc-pVTZ level of theory. The automatic methods in Gaussian were used to compute the Raman activities (R_A) for $90°$ scattering

$$R_A = (45(\alpha')^2_k + 7(\gamma')^2_k)/45 \tag{4.2.29}$$

where $(\alpha')^2_k$ is a square magnitude of derivative of the isotropic part of the polarizability tensor with respect to the k-th normal mode, and $(\gamma')^2_k$ is a square magnitude of derivative of the anisotropic part of the polarizability tensor with respect to the k-th normal mode and the derivatives are evaluated at zero displacement. It should be noted that R_A is proportional to the Raman intensity I. The calculations of R_A [101] show that for all vibrational modes of the CH_4–N_2 complex the values of

R_A are decreased by factors 0.7–0.98 in comparison with R_A for the vibrational modes of the monomers CH_4 and N_2.

As to the depolarization ratio ρ, which describes the polarization properties of Raman scattering, the calculations of ρ [101] have shown that the polarization properties of Raman scattering are retained in the complex for vibrational frequencies corresponding to the frequencies of monomers CH_4 and N_2 (anisotropic Raman scattering: v_2, v_3 and v_4 of CH_4 molecule and isotropic Raman scattering: v_1 and v of CH_4 and N_2). The calculations have also shown that the Raman scattering for intermolecular vibrations (the harmonic modes 19.6, 23.3, 65.7 and 71.8 cm^{-1}) has purely anisotropic nature ($\rho \approx 3/4$ and $\rho \approx 6/7$ for linearly polarized and unpolarized incident light, respectively). However, for the vibrational mode of 48.2 cm^{-1} the Raman scattering is the mixture of the anisotropic and isotropic scattering. It should be noted that for collisional Raman scattering the Raman line intensities of the CH_4–N_2 complex could be larger than predicted ones for his configuration 4.

4.2.5   Polarizabilities of C_2H_4–C_2H_4 Complex

As was mention in Sect. 3.2.4, ethylene, in spite of its chemical simplicity, is of much interest because takes place in many diverse processes. Even in biology it is of great interest, for example, ethylene is like a hormone that regulates a number of physiological processes in the plants [102]. Nonbonding interactions of π-electron systems have been intensively studied, since π-π interactions control several phenomena such as crystal packing of unsaturated hydrocarbon molecules, conformational preference of nucleic acids, and host-guest interactions of aromatic molecules. Moreover, ethylene dimer as long as the methane-nitrogen complex is of particular interest for astrophysical applications. These complexes exist in the atmospheres of giant planets (Jupiter [103, 104], Saturn [105], Neptune and Uranus [106]) and Saturn's satellite Titan [107–111]. The ethylene dimer is the simplest π-π organic complex that can serve as a test object for nonbonding interaction theories.

In this Section the results of polarizability calculations for the 12 configurations of the C_2H_4–C_2H_4 complex are given (Table 4.1). The input parameters of the configurations for calculations are the same as described in Sect. 3.2.4. The value of applied electric homogeneous field used in the finite-field procedure was chosen to be 0.001 a.u.

In this way the following values of the polarizability components of the separate C_2H_4 molecule were also obtained (in a.u.): $\alpha_{xx} = 22.05$, $\alpha_{yy} = 24.96$, $\alpha_{zz} = 34.24$ which agree with the experimental ones $\alpha_{xx} = 22.94$, $\alpha_{yy} = 26.04$, $\alpha_{zz} = 36.44$ [20].

Table 4.1 Calculated polarizabilities (in a.u.) of C_2H_4–C_2H_4 complex at the CCSD(T)/ aug-cc-pVTZ level of theory with the BSSE correction

R, Å	Config.	α_{xx}	α_{yy}	α_{zz}	Config.	α_{xx}	α_{yy}	α_{zz}
3.5	1	40.12	43.43	90.02	7	46.25	52.18	62.13
4		41.32	45.62	80.91		47.25	53.23	61.35
4.5		42.32	47.24	77.41		47.97	54.04	60.56
5		42.91	48.16	75.23		48.47	54.61	59.78
5.5		43.25	48.67	73.56		48.82	55.01	59.23
6		43.49	49.02	72.35		49.06	55.29	58.62
7		43.80	49.45	70.89		49.38	55.65	57.63
8		43.96	49.68	70.11		49.56	55.86	57.18
9		44.06	49.81	69.66		49.66	55.99	56.91
10		44.12	49.89	69.38		49.73	56.07	56.75
3.5	2	52.31	52.31	54.84	8	42.68	42.68	87.36
4		53.38	53.38	54.18		43.85	43.85	80.15
4.5		54.12	54.12	53.33		44.74	44.74	76.90
5		54.65	54.65	52.57		45.37	45.37	74.85
5.5		55.02	55.02	51.97		45.80	45.80	73.30
6		55.30	55.30	51.52		46.10	46.10	72.15
7		55.65	55.65	50.95		46.46	46.46	70.71
8		55.86	55.86	50.62		46.65	46.65	69.93
9		55.98	55.98	50.41		46.76	46.76	69.48
10		56.07	56.07	50.28		46.83	46.83	69.20
3.5	3	47.23	63.34	46.88	9	51.79	43.12	67.98
4		47.94	64.67	46.79		52.99	44.28	66.08
4.5		48.43	65.61	46.45		53.85	45.05	64.56
5		48.78	66.28	46.02		54.47	45.58	63.30
5.5		49.03	66.76	45.62		54.90	45.93	62.33
6		49.21	67.11	45.30		55.21	46.18	61.61
7		49.45	67.58	44.87		55.61	46.49	60.70
8		49.60	67.86	44.62		55.84	46.67	60.20
9		49.69	68.04	44.47		55.97	46.77	59.90
10		49.75	68.15	44.37		56.06	46.84	59.72
3.5	4	55.52	55.52	46.90	10	53.34	40.72	69.88
4		56.47	56.47	46.78		55.02	41.66	66.97
4.5		57.14	57.14	46.44		56.21	42.35	64.9
5		57.62	57.62	46.01		57.00	42.83	63.49
5.5		57.96	57.96	45.61		57.55	43.15	62.39
6		58.21	58.21	45.30		57.92	43.38	61.62
7		58.55	58.55	44.87		58.40	43.65	60.69
8		58.75	58.75	44.62		58.67	43.80	60.20
9		58.88	58.88	44.47		58.83	43.89	59.90

(continued)

Table 4.1 (continued)

R, Å	Config.	α_{xx}	α_{yy}	α_{zz}	Config.	α_{xx}	α_{yy}	α_{zz}
10		58.96	58.96	44.37		58.93	43.95	59.71
3.5	5	43.99	54.04	62.81	11	63.03	44.32	52.01
4		44.81	55.44	61.71		64.44	45.06	51.06
4.5		45.40	56.47	60.87		65.45	45.57	50.21
5		45.81	57.18	60.05		66.16	45.92	49.48
5.5		46.10	57.67	59.30		66.67	46.16	48.91
6		46.30	58.01	58.68		67.04	46.34	48.48
7		46.55	58.45	57.86		67.54	46.58	47.94
8		46.70	58.69	57.40		67.84	46.71	47.63
9		46.79	58.84	57.13		68.02	46.80	47.45
10		46.85	58.94	56.96		68.14	46.85	47.34
3.5	6	55.37	52.59	51.38	12	62.25	41.04	57.07
4		56.34	53.58	50.81		63.95	41.94	55.19
4.5		57.04	54.28	50.11		65.13	42.55	53.81
5		57.54	54.76	49.44		66.11	43.11	52.94
5.5		57.90	55.11	48.89		66.71	43.42	52.26
6		58.17	55.36	48.48		67.12	43.61	51.75
7		58.52	55.68	47.94		67.67	43.85	51.13
8		58.74	55.88	47.63		67.99	43.99	50.79
9		58.87	56.00	47.45		68.18	44.08	50.59
10		58.96	56.07	47.34		68.31	44.13	50.46

References

1. S. Kielich, *Molekularna Optyka Nieliniowa (Nonlinear Molecular Optics)* (Naukowe, Warszawa-Poznan, 1977)
2. H.J. Achtermann, G. Magnus, T.K. Bose, Refractivity virial coefficients of gaseous CH_4, C_2H_4, C_2H_6, CO_2, SF_6, H_2, N_2, He and Ar. J. Chem. Phys. **94**(8), 5669–5684 (1991)
3. U. Hohm, K. Kerl, A Michelson twin interferometer for precise measurements of refractive index of gases between 100 K and 1300 K. Meas. Sci. Technol. **1**(4), 329–336 (1990)
4. U. Hohm, K. Kerl, Interferometric measurements of dipole polarizability α; of molecules between 300 K and 1100 K. II. A new method measuring the dispersion of the polarizability and its application to Ar, H_2, and O_2. Mol. Phys. **69**(5), 819–831 (1990)
5. K. Kerl, U. Hohm, H. Varchmin, Polarizability $\alpha(\omega, T, \rho)$ of small molecules in the gas phase. Ber. Bunsenges. Phys. Chem. **96**(5), 728–733 (1992)
6. U. Hohm, Frequency-dependence of second refractivity virial coefficients of small molecules between 325 nm and 633 nm. Mol. Phys. **81**(1), 157–168 (1994)
7. M. Monan, J.-L. Bribes, R. Gaufres, Measurement of weak depolarization ratios of the Rayleigh scattering of gases using Raman spectroscopy. J. Raman Spectroscopy **12**(2), 190–193 (1982)
8. X. Song, J. Jonas, T.W. Zerda, Total intensity measurements of depolarized Rayleigh scattering from linear molecules. J. Phys. Chem. **93**(19), 6887–6890 (1989)

9. A.D. Buckingham, M.P. Bogaard, D.A. Dunmur, C.P. Hobbs, B.J. Orr, Kerr effect in the some simple non-dipolar gases.` Trans. Faraday Soc. **66**(7), 1548–1553 (1970)

10. J.H. Williams, Dispersion studies in the Kerr effect of molecular oxygen. Chem. Phys. Lett. **147**(6), 585–590 (1988)

11. S. Gustafson, W. Gordy, The microwave Stark effect in oxygen. Phys. Lett. **49**(2), 161–162 (1974)

12. T.E. Gough, B.J. Orr, G. Scoles, Laser Stark spectroscopy of carbon dioxide in a molecular beam. J. Mol. Spectr. **99**(1), 143–158 (1983)

13. W.Q. Cai, T.E. Gough, X.J. Gu, N.R. Isenor, G. Scoles, Polarizability of CO_2 studied in molecular beam laser Stark spectroscopy. Phys. Rev. A **36**(10), 4722–4727 (1987)

14. M.J. Dyer, W.K. Bischel, Optical Stark shift spectroscopy. Measurement of the v = 1 polarizability in H_2. Phys. Rev. A **44**(5), 3138–3143 (1991)

15. A.N. Vereshchagin, *The Polarizability of Molecules* (Nauka, Moscow, 1980). (in Russian)

16. G.A. Vandysheva, V.N. Savel'ev, L.N. Sinitsa, Investigation of absorption spectrum for the "2-0" transition induced by electric field. II. Determination of the matrix elements of the polarizability tensor. Atmos. Ocean. Opt. **3**(4), 360–363 (1990)

17. G. Placzek, in *Handbuch der Radiologie*, ed. by E. Marx, Teir 2, vol. VI (Akademische Verlagsgesselschaft, Leipzig, 1934), pp. 205–374

18. W.F. Murphy, W. Holzer, H.J. Bernstein, Gas phase raman intensities: a review of "pre-laser" data. Appl. Spectrosc. **23**(2), 211–218 (1969)

19. H. Hettema, P.E.S. Wormer, P. Jorgensen, H.J.A. Jensen, T. Helgaker, Frequency-dependent polarizabilities of O_2 and van der Waals coefficients of dimers containing O_2. J. Chem. Phys. **100**(2), 1297–1302 (1994)

20. A. Weber, S. Brodersen, J.M. Friedman, H.W. Klöckner, G.V.knighten, J.W. Nibler, D.L. Rousseau, H.W. Schrötter, R.P. Srivastava, A. Weber, P.F. Williams, H.R. Zaidi (eds.), *Raman Spectroscopy of Gases and Liquids* (Springer, Berlin, 1979)

21. M.A. Buldakov, I.I. Matrosov, T.N. Popova, Determination of the anisotropy for polarizability tensor of N_2 and O_2 molecules. Opt. Spektrosk **46**, 867–869 (1979). (in Russian)

22. R.L. Rowell, G.M. Aval, J.J. Barrett, Rayleigh-Raman depolarization of light scattered by gases. J. Chem. Phys. **54**(5), 1960–1964 (1971)

23. M. Monan, J.-L. Bribes, R. Gaufres, Measurement of weak depolarization ratios of the Rayleigh scattering of gases using Raman spectroscopy. J. Raman Spectrosc. **12**(2), 190–193 (1982)

24. G.R. Alms, A.K. Burnham, W.H. Flygare, Measurement of the dispersion in polarizability anisotropies. J. Chem. Phys. **63**(8), 3321–3326 (1975)

25. F. Baas, K.D. van den Hout, Measurement of depolarization ratios and polarizability anisotropies of gaseous molecules. Phys. A **95**(3), 597–601 (1979)

26. W.F. Murphy, The Rayleigh depolarization ratio and rotational Raman spectrum of water vapor and the polarizability components for the water molecule. J. Chem. Phys. **67**(12), 5877–5882 (1977)

27. M.P. Bogaard, A.D. Buckingham, R.K. Pierens, A.N. White, Rayleigh scattering depolarization ratio and molecular polarizability anisotropy for gases. J. Chem. Soc. Faraday Trans. **74**(12), 3008–3015 (1978)

28. N.J. Bridge, A.D. Buckingham, The polarization of laser light scattered by gases. Proc. Roy. Soc. A **295**(1442), 334–349 (1966)

29. N.J. Bridge, A.D. Buckingham, Polarization of laser light scattered by gases. J. Chem. Phys. **40**(9), 2733–2734 (1964)

30. M.A. Buldakov, I.I. Ippolitov, B.V. Korolev, I.I. Matrosov, A.E. Cheglokov, V.N. Cherepanov, YuS Makushkin, O.N. Ulenikov, Vibration rotation Raman spectroscopy of gas media. Spectrochim. Acta A **52**(8), 995–1007 (1996)

31. J.M. Hoell, F. Allario, O. Jarret, R.K. Seals, Measurements of F_2, NO and ONF Raman cross section and depolarization ratios for diagnostics in chemical lasers. J. Chem. Phys. **58**(7), 2896–2901 (1973)

32. W. Holzer, Y. Le Duff, The depolarization ratio of the Raman bands of the vibration of diatomic molecules. in *Adv. Raman Specrosc*, vol. 1 (Heyden, London, 1973), pp. 109–112

33. C.M. Penney, L.M. Goldman, M. Lapp, Raman scattering cross sections. Nat. Phys. Sci. **235** (7), 110–112 (1972)

34. T. Yoshino, H. Bernstein, Intensity in the Raman effect. 6. The photoelectrically recorded Raman spectra of some gases. J. Molec Spectrosc. **2**(3), 213–240 (1958)

35. W. Knippers, K. Van Helvoort, S. Stolte, Vibrational overtones of the homonuclear diatomics (N_2, O_2, D_2) observed by the spontaneous Raman effect. Chem. Phys. Lett. **121** (4,5), 279–286 (1985)

36. C.M. Penney, R.L.S. Peters, M. Lapp, Absolute rotational Raman cross sections for N_2, O_2, and CO_2. J. Opt. Soc. Am. **64**(5), 712–716 (1974)

37. W.R. Fenner, H.A. Hyatt, J.M. Kellam, S.P.S. Porto, Raman cross section of some simple gases. J. Opt. Soc. Am. **63**(1), 73–77 (1973)

38. H.A. Hyatt, J.M. Cherlow, W.R. Fenner, S.P.S. Porto, Cross section for the Raman effect in molecular nitrogen gas. J. Opt. Soc. Am. **63**(12), 1604–1606 (1973)

39. A.D. Buckingham, A. Szabo, Determination of derivatives of the polarizability anisotropy in a diatomic molecule from relative Raman intensities. J. Raman Spectrosc. **7**(1), 46–48 (1978)

40. U. Hohm, Experimental static dipole-dipole polarizabilities of molecules. J. Mol. Struct. **1054–1055**, 282–292 (2013)

41. G. Maroulis, A systematic study of basis set, electron correlation, and geometry effects on the electric multipole moments, polarizability, and hyperpolarizability of HCl. J. Chem. Phys. **108**(13), 5432–5448 (1998)

42. A.D. Buckingham, The polarizability of a pair of interacting atoms. Trans. Faraday Soc. **52**, 1035–1041 (1956)

43. A.D. Buckingham, K.L. Klarke, Long-range effects of molecular interactions on the polarizability of atoms. Chem. Phys. Lett. **57**(3), 321–325 (1978)

44. K.L.C. Hunt, B.A. Zilles, J.E. Bohr, Effect of van der Waals interactions on the polarizability of atoms, oscillators, and dipolar rotors at long range. J. Chem. Phys. **75**(6), 3079–3086 (1981)

45. K.L.C. Hunt, Y.Q. Liang, S. Sethuraman, Transient, collision-induced changes in polarizability for atoms interacting with linear, centrosymmetric molecules at long range. J. Chem. Phys. **89**(12), 7126–7138 (1988)

46. D.M. Bishop, J. Pipin, Calculation of the polarizability and hyperpolarizability tensors, at imaginary frequency, for H, He, and H_2 and the dispersion polarizability coefficients for interactions between them. J. Chem. Phys. **97**(1), 3375–3381 (1992)

47. P.W. Fowler, K.L.C. Hunt, H.M. Kelly, A.J. Sadley, Multipole polarizabilities of the helium atom and collision-induced polarizabilities of pairs containing He or H atoms. J. Chem. Phys. **100**(4), 2932–2935 (1994)

48. X. Li, K.L.C. Hunt, Transient changes in polarizability for centrosymmetric linear molecules interacting at long range: theory and numerical results for H_2-H_2, H_2-N_2, and N_2-N_2. J. Chem. Phys. **100**(11), 7875–7889 (1994)

49. C. Domene, P.W. Fowler, P. Jemmer, P.A. Madden, Dipole-induced-dipole polarizabilities of symmetric clusters. Mol. Phys. **98**(18), 1391–1407 (2000)

50. M.A. Buldakov, B.V. Korolev, I.I. Matrosov, V.N. Cherepanov, Polarizability of two interacting molecules N_2 and O_2. Opt. Spectrosc. **94**, 185–190 (2003)

51. M.A. Buldakov, V.N. Cherepanov, N.S. Nagornova, Polarizability of the van der Waals complexes N_2...Y and O_2...Y (Y = He, Ne, Ar, Kr, Xe). Part 1. Stable configurations, Atmos. Ocean. Opt. **18**(1–2), 12–17 (2005)

52. M.A. Buldakov, V.N. Cherepanov, N.S. Nagornova, Polarizability of the van der Waals complexes N_2...Y and O_2...Y (Y = He, Ne, Ar, Kr, Xe). Part 2. Unstable configurations, Atmos. Ocean. Opt. **18**(1–2), 18–22 (2005)

53. M.A. Buldakov, V.N. Cherepanov, N.S. Nagornova, Polarizability of the van der Waals complexes N_2...Y and O_2...Y (Y = He, Ne, Ar, Kr, Xe). Part 3. Frequency dependence, Atmos. Ocean. Opt. **19**(1), 33–37 (2006)

54. X. Li, C. Ahuja, J.F. Harrison, K.L.C. Hunt, The collision-induced polarizability of a pair of hydrogen molecules. J. Chem. Phys. **126**(21), 214302 (2007)
55. A.D. Buckingham, in *Intermolecular Interactions: From Diatomics to Biopolymers* (Wiley, New York, 1978)
56. P.W. Fowler, K.L.C. Hunt, H.M. Kelly, A.J. Sadley, Multipole polarizabilities of the helium atom and collision-induced polarizabilities of pairs containing He and H atoms. J. Chem. Phys. **100**, 2932–2935 (1994)
57. S.M. El-Sheikh, G.C. Tabisz, A.D. Buckingham, Collision-induced light scattering by isotropic molecules: the role of the quadrupole polarizability. Chem. Phys. **247**, 407–412 (1999)
58. K.L.C. Hunt, J.E. Bohr, Field-induced fluctuation correlations and the effects of van der Waals interactions on molecular polarizabilities. J. Chem. Phys. **84**, 6141–6150 (1986)
59. M.A. Buldakov, V.N. Cherepanov, Asymptotic model of exchange interactions for polarizability calculation of van der Waals. J. Comp. Meth. Sci. Eng. **10**(3–6), 165–181 (2010)
60. K.L.C. Hunt, A.D. Buckingham, The polarizability of H_2 in the triplet state. J. Chem. Phys. **72**, 2832–2840 (1980)
61. B. Fernandez, C. Hättig, H. Koch, A. Rizzo, *Ab initio* calculation of the frequency-dependent interaction induced hyperpolarizability of Ar_2. J. Chem. Phys. **110**, 2872–2882 (1999)
62. C. Hättig, H. Larsen, J. Olsen, P. Jørgensen, H. Koch, B. Fernandez, A. Rizzo, The effect of intermolecular interactions of the electric properties of helium and argon. I. *Ab initio* calculation of the interaction induced polarizability in He_2 and Ar_2. J. Chem. Phys. **111**, 10099–10107 (1999)
63. L. Silberstein, Molecular refractivity and atomic interaction. Phil. Mag. **33**(193), 92–128 (1917); Phil. Mag. **33**(198), 521–533 (1917)
64. K. Patel, P.R. Butler, A.M. Ellis, M.D. Wheeler, *Ab initio* study of Rg–N_2 and Rg–C_2 van der Waals complexes (Rg = He, Ne, Ar). J. Chem. Phys. **119**(2), 909–920 (2003)
65. C.-H. Hu, A.J. Thakkar, Potential energy for interaction between N_2 and He: *Ab initio* calculations, analytic fits, and second virial coefficients. J. Chem. Phys. **104**(7), 2541–2547 (1996)
66. M.S. Bowers, K.T. Tang, J.P. Toennies, The anisotropic potentials of He–N_2, Ne–N_2 and Ar–N_2. J. Chem. Phys. **88**(9), 5465–5474 (1988)
67. L. Beneventi, P. Casavecchia, G.G. Volpi, C.C.K. Wong, F.R.W. McCourt, G.C. Corey, D. Lemoine, On the N_2–He potential energy surface. J. Chem. Phys. **95**(8), 5827–5845 (1991)
68. M.C. Salazar, J.L. Paz, A.J. Hernandez, *Ab initio* test study of the N_2...H_2 and N_2...He van der Waals dimers. J. Mol. Struct.: TEOCHEM. **464**(1–3), 183–189 (1999)
69. A.K. Dham, W.J. Meath, Exchange-Coulomb potential energy surfaces and related physical properties for Ne-N_2. Mol. Phys. **99**(12), 991–1004 (2001)
70. W. Jager, Y. Xu, G. Armstrong, M.C.L. Gerry, F.Y. Naumkin, F. Wang, F.R.W. McCourt, Microwave spectra of the Ne-N_2 Van der Waals complex: experiment and theory. J. Chem. Phys. **109**(13), 5420–5432 (1998)
71. B. Fernandez, H. Koch, J. Makarewicz, Accurate intermolecular ground state potential of the Ar-N_2 complex. J. Chem. Phys. **110**(17), 8525–8532 (1999)
72. W. Jaeger, M.C.L. Gerry, C. Bissonnette, F.R.W. McCourt, Pure rotational spectrum of, and potential-energy surface for the Ar-N_2 van der Waals complex. Faraday Discuss. **97**, 105–118 (1994)
73. A.K. Dham, F.R.W. McCourt, W.J. Meath, Exhange-Coulomb model potential energy surface for the N_2-Ar interaction. J. Chem. Phys. **103**(19), 8477–8491 (1995)
74. F.Y. Naumkin, Molecular versus atom-atom interaction anisotropy in the case of the Ar-N_2 van der Waals system. Mol. Phys. **90**(6), 875–888 (1997)
75. L. Beneventi, P. Casavecchia, G.G. Volpi, C.C.K. Wong, F.R.W. McCourt, Multiproperty determination of a new N_2-Ar intermolecular interaction potential energy surface. J. Chem. Phys. **98**(10), 7926–7939 (1993)

76. M.A. Ter Horst, C.J. Jameson, Classical trajectories on simple model potentials and other data. J. Chem. Phys. **102**(11), 4431–4446 (1995)

77. A.C. De Dios, C.J. Jameson, The ^{129}Xe nuclear shielding surfaces for Xe interacting with linear molecules CO_2, N2 and CO. J. Chem. Phys. **107**(11), 4253–4270 (1997)

78. G.C. Groenenboom, I.M. Struniewicz, Three-dimensional ab initio potential energy surface for He-O_2. J. Chem. Phys. **113**(21), 9562–9566 (2000)

79. S.M. Cybulski, R. Burcl, M.M. Szczesniak, G. Chalasinski, *Ab initio* study of the $O_2(X^3\Sigma_g^-)$ + He (^1S) van der Waals cluster. J. Chem. Phys. **104**(20), 7997–8002 (1996)

80. L. Beneventi, P. Casavecchia, F. Pirani, F. Vecchiocattivi, G.G. Volpi, G. Brocks, A. van der Avoird, B. Heijmen, J. Reuss, The Ne-O_2 potential energy surface from high-resolution diffraction and glory scattering experiments and from the Zeeman spectrum. J. Chem. Phys. **95**(1), 195–204 (1991)

81. S.M. Cybulski, R.A. Kendall, G. Chalasinski, M.W. Severson, M.M. Szczesniak, *Ab initio* study of the $O_2(X^3\Sigma_g^-)$ + Ar (^1S) van der Waals interaction. J. Chem. Phys. **106**(18), 7731–7737 (1997)

82. V. Aquilanti, D. Ascenzi, D. Cappelletti, M. De Castro, F. Pirani, Scattering of aligned molecules. The potential energy surfaces for the Kr-O_2 and Xe-O_2 systems. J. Chem. Phys. **109**(10), 3898–3910 (1998)

83. M.A. Buldakov, V.N. Cherepanov, Electronic polarizability of the N_2 and O_2 molecules: the role of exchange interactions. Atmos. Ocean. Opt. J. **16**(11), 928–932 (2003)

84. D. Spelsberg, W. Meyer, *Ab initio* dynamic dipole polarizabilities for O_2, its photoabsorption spectrum in the Schumann—Runge region, and long-range interaction coefficients for its dimer. J. Chem. Phys. **109**(22), 9802–9810 (1998)

85. D. Spelsberg, W. Meyer, Dynamic multipole polarizabilities, reduced spectra, and interaction coefficients for N_2 and CO. J. Chem. Phys. **111**(21), 9618–9624 (1999)

86. A.J. Thakkar, H. Hettema, P.E.S. Wormer, *Ab initio* dispertion coefficients for interactions involving rare-gas atoms. J. Chem. Phys. **97**(5), 3252–3257 (1992)

87. M.N. Adamov, YuB Malykhanov, V.V. Meshkov, R.M. Chadin, Calculation of optical characteristics of atoms with a closed shell by the Hartree-Fock-Roothaan method. Opt. Spectrosc. **96**(2), 192–194 (2004)

88. M. Masili, A.F. Starace, The interaction polarizability of two nitrogen molecules. Phys. Rev. A **68**, 012508 (2003)

89. H. Naus, W. Ubach, Visible absorption bands of the $(O_2)_2$ collision complex at pressures below 760 Torr. Appl. Opt. **38**(15), 3423–3428 (1999)

90. B. Brunetti, G. Liuti, E. Luzzatti, F. Pirani, F. Vecchiocattivi, Study of the interactions of atomic and molecular oxygen with O_2 and N_2 by scattering data. J. Chern. Phys. **74**(12), 6734–6741 (1981)

91. D.G. Bounds, The interaction polarizability of two hydrogen molecules. Mol. Phys. **38**(6), 2099–2106 (1979)

92. D.G. Bounds, A. Hinchliffe, C.J. Spicer, The interaction polarizability of two hydrogen molecules. Mol. Phys. **42**(1), 73–82 (1981)

93. M. Bartolomei, E. Carmona-Novillo, M.I. Hernández, J. Campos-Martínez, R. Hernández-Lamoneda, Long-range interaction for dimers of atmospheric interest: dispersion, induction and electrostatic contributions for O_2-O_2, N_2-N_2 and O_2-N_2. J. Comput. Chem. **32**(2), 279–290 (2011)

94. D. Cappelletti, F. Pirani, B. Bussery-Honvault, L. Gomezc, M. Bartolomei, A bond–bond description of the intermolecular interaction energy: the case of weakly bound N2–H2 and N_2–N_2 complexes. Phys. Chem. Chem. Phys. **10**, 4281–4293 (2008)

95. J.M. Fernández, G. Tejeda, A. Ramos B.J. Howard, S. Montero, Gas-Phase Raman Spectrum of NO Dimer. J. Mol. Spectr. **194**, 278–280 (1999)

96. M.A. Buldakov, B.V. Korolev, I.I. Matrosov, T.N. Popova, Polarizability of the N_2 and O_2 molecules. Opt. Spectrosk. **62**(3), 519–523 (1987). (USSR)

97. M.A. Buldakov, V.N. Cherepanov, YuN Kalugina, N. Zvereva-Loëte, V. Boudon, Static polarizability surfaces of the van der Waals complex CH_4–N_2. J. Chem. Phys. **132**(16), 164304 (2010)

98. B. D'Acunto, in *Computational Methods for PDE in Mechanics* (Chap. Finite differences). Series on Advances in Mathematics for Applied Sciences 67 (World Scientific, London, 2004)
99. G. Maroulis, Electric dipole hyperpolarizability and quadrupole polarizability of methane from finite-field coupled cluster and fourth-order many-body perturbation theory calculations. Chem. Phys. Lett. **226**(3–4), 420–426 (1994)
100. G. Maroulis, Accurate electric multipole moment, static polarizability and hyperpolarizability derivatives for N_2. J. Chem. Phys. **118**(6), 2673–2687 (2003)
101. Y.N. Kalugina, Theoretical investigation of the potential energy, dipole moment and polarizability surfaces of the CH_4–N_2 and C_2H_4–C_2H_4 van der Waals complexes, Thesis to obtained the degree of Doctor of Physics (ICB, Dijon, 2010), p. 206
102. B.M. Binder, The ethylene receptors: complex perception for a simple gas. Plant Sci. **175**, 8–17 (2008)
103. J.I. Moses, T. Fouchet, R.V. Yelle, A.J. Friedson, G.S. Orton, B. Bézard, F. Drossart, G.R. Gladstone, T. Kostiuk, T.A. Livengood, The stratosphere of Jupiter. in *Jupiter: The Planet, Satellites, and Magnetosphere*, ed. by F. Bagenal, T.E. Dowling, W.B. McKinnon (Cambridge University Press, New York, 2004), pp. 129–158
104. G.P. Smith, D. Nash, Local sensitivity analysis for observed hydrocarbons in a Jupiter photochemistry model. Icarus **182**(1), 181–201 (2006)
105. R. Prangé, T. Fouchet, R. Courtin, J.E.P. Connerney, J.C. McConnell, Latitudinal variation of Saturn photochemistry deduced from spatially-resolved ultraviolet spectra. Icarus **180**(2), 379–392 (2006)
106. J.I. Lunine, The atmospheres of Uranus and Neptune. Annu. Rev. Astron. Astrophys. **31**, 217–263 (1993)
107. A. Coustenis, F.W. Taylor (eds.), *Titan: exploring an earthlike world* (World Scientific Publishing Co., Pte. Ltd., 2008), p. 412
108. A. Coustenis, R.K. Achterberg, B.J. Conrath, D.E. Jennings, A. Marten, D. Gautier, C.A. Nixon, F.M. Flasar, N.A. Teanby, B. Bézard, R.E. Samuelson, R.C. Carlson, E. Lellouch, G. L. Bjoraker, P.N. Romani, F.W. Taylor, P.G.J. Irwin, T. Fouchet, A. Hubert, G.S. Orton, V. G. Kunde, S. Vinatier, J. Mondellini, M.M. Abbas, R. Courtin, The composition of Titan's stratosphere from Cassini/CIRS mid-infrared spectra. Icarus **189**(1), 35–62 (2007)
109. J. Cui, R.V. Yelle, V. Vuittonb, J.H. Waite Jr., W.T. Kasprzak, D.A. Gell, H.B. Niemannd, I. C.F. Muller-Wodarg, N. Borggren, G.G. Fletcher, E.L. Patrick, E. Raaen, B.A. Mageec, Analysis of Titan's neutral upper atmosphere from cassini ion neutral mass spectrometer measurements. Icarus **200**(2), 581–615 (2009)
110. H.G. Roe, I. de Pater, C.P. McKay, Seasonal variation of Titan's stratospheric ethylene (C_2H_4) observed. Icarus **169**(2), 440–461 (2004)
111. V. Vuitton, J.-F. Doussin, Y. Benilan, F. Raulin, M.-C. Gazeau, Experimental and theoretical study of hydrocarbon photochemistry applied to Titan stratosphere. Icarus **185**(1), 287–300 (2006)
112. G. Maroulis, A. Haskopoulos, D. Xenides, New basis sets for xenon and interaction polarizability of two xenon atoms. Chem. Phys. Lett. **396**, 59–95 (2004)
113. M.J. Frisch, G.W. Trucks, H.B. Schlegel, G.E. Scuseria, M.A. Robb, J.R. Cheeseman, J.A. Montgomery, Jr., T. Vreven, K.N. Kudin, J.C. Burant, J.M. Millam, S.S. Iyengar, J. Tomasi, V. Barone, B. Mennucci, M. Cossi, G. Scalmani, N. Rega, G.A. Petersson, H. Nakatsuji, M. Hada, M. Ehara, K. Toyota, R. Fukuda, J. Hasegawa, M. Ishida, T. Nakajima, Y. Honda, O. Kitao, H. Nakai, M. Klene, X. Li, J.E. Knox, H.P. Hratchian, J.B. Cross, V. Bakken, C. Adamo, J. Jaramillo, R. Gomperts, R.E. Stratmann, O. Yazyev, A.J. Austin, R. Cammi, C. Pomelli, J.W. Ochterski, P.Y. Ayala, K. Morokuma, G.A. Voth, P. Salvador, J.J. Dannenberg, V.G. Zakrzewski, S. Dapprich, A.D. Daniels, M.C. Strain, O. Farkas, D.K. Malick, A.D. Rabuck, K. Raghavachari, J.B. Foresman, J.V. Ortiz, Q. Cui, A.G. Baboul, S. Cliord, J. Cioslowski, B.B. Stefanov, G. Liu, A. Liashenko, P. Piskorz, I. Komaromi, R.L. Martin, D.J. Fox, T. Keith, M.A.

Al-Laham, C.Y. Peng, A. Nanayakkara, M. Challacombe, P.M.W. Gill, B. Johnson, W. Chen, M.W. Wong, C. Gonzalez, J.A. Pople, in *Gaussian 03, Revision D.02* (Gaussian, Inc., Wallingford CT, 2004)

Chapter 5
Interaction-induced Hyperpolarizability

At present, in spite of the well-known fact that the interaction of atoms and molecules leads to the changing of their multipole moments and (hyper)polarizabilities [1–3], only the simple moments and polarizabilities of interacting molecules such as interaction-induced dipole moments and dipole polarizabilities have widely studied. Nevertheless, the induced dipole moments and polarizabilities play the important role in understanding the collision-induced absorption, collision-induced Rayleigh scattering or collision-induced Raman light scattering [4–9].

Modern calculations of the hyperpolarizabilities of interacting atom-molecular systems are based on the ab initio methods and analytical models of long-range multipole-induced multipole interactions. These methods have both advantages and disadvantages. Analytical models are only effective for interacting systems at large intermolecular separations, the exchange and other overlap effects can be ignored. For these separations the hyperpolarizability surfaces can be calculated fully and correctly if multipole moments, polarizabilities and higher polarizabilities of the interacting molecules (or atoms) are known. Now, the analytical models are applied, as a rule, for binary complexes, which are composed of atoms and simple molecules [10–15].

Ab initio methods, in their turn, allow calculating the hyperpolarizabilities of interacting systems both for small and large separations. However, when a lot of points on the (hyper)polarizability surfaces (greed with a small step) are needed the use of high level ab initio methods becomes often not effective in comparison with analytical calculations. Such problems can appear, for example, in description of collision processes. Now, computer methods are widely and effectively used for calculation of the hyperpolarizabilities both for small atom-molecular systems and for more complicated atom-molecular systems [16–35] and can be applied to study nonlinear scattering effects. For instant, there is a lot of publications devoted to the study of collision-induced hyper-Rayleigh spectra for binary mixtures of noble gases [36–40], H_2–Ar and H_2–He gas mixtures [41, 42], and gaseous SF_6 [43].

© The Author(s) 2017
V.N. Cherepanov et al., *Interaction-induced Electric Properties of van der Waals Complexes*,
SpringerBriefs in Electrical and Magnetic Properties of Atoms, Molecules, and Clusters,
DOI 10.1007/978-3-319-49032-8_5

5.1 Theoretical Treatment

5.1.1 Computational Details

The components of the first hyperpolarizability tensor of the complex are usually calculated employing the finite-field method described above for the dipole moment and polarizability. Using this method, the first hyperpolarizability is determined using Eq. (2.4.2) or the third derivative of the energy over the external electric field:

$$\beta_{\alpha\beta\gamma} = -\left(\frac{\partial^3 E}{\partial F_\alpha^0 \partial F_\beta^0 \partial F_\gamma^0}\right)_{\mathbf{F}^0=0}. \tag{5.1.1}$$

As described above, in the framework of this method one can derive the formulae for calculation of the first hyperpolarizability as follows:

$$\beta_{\alpha\alpha\alpha} = -\frac{E(2F_\alpha) - E(-2F_\alpha) - 2E(F_\alpha) + 2E(-F_\alpha)}{2F_\alpha^3}, \tag{5.1.2}$$

$$\beta_{\alpha\alpha\beta} = -\frac{E(F_\alpha, F_\beta) - E(F_\alpha, -F_\beta) + E(-F_\alpha, F_\beta) - E(-F_\alpha, -F_\beta) - 2E(F_\beta) + 2E(-F_\beta)}{2F_\alpha^2 F_\beta}, \tag{5.1.3}$$

$$\beta_{\alpha\beta\gamma} = -\frac{E(-F_\alpha, -F_\beta, F_\gamma) - E(-F_\alpha, -F_\beta, -F_\gamma) + E(F_\alpha, F_\beta, F_\gamma) - E(F_\alpha, F_\beta, -F_\gamma) - E(F_\alpha, -F_\beta, F_\gamma)}{8F_\alpha F_\beta F_\gamma}$$
$$-\frac{E(F_\alpha, -F_\beta, -F_\gamma) - E(-F_\alpha, F_\beta, F_\gamma) + E(-F_\alpha, F_\beta, -F_\gamma)}{8F_\alpha F_\beta F_\gamma}. \tag{5.1.4}$$

In this way, the higher polarizabilities give some contributions to $\beta_{\alpha\beta\gamma}$. However, the use of the procedure described in Sect. 3.1.1 helps to remove these contributions. Thus, G. Maroulis [44] has obtained the following expressions for $\beta_{\alpha\beta\gamma}$:

$$\beta_{zzz} = (-64D_z(F) + 34D_z(2F) - D_z(4F))/(24F^3), \tag{5.1.5}$$

$$\beta_{zxx} = (32D_{xz}(F) - 34D_{xz}(2F))/(12F^3) \tag{5.1.6}$$

where

$$D_z(F) = (E(-F_z) - E(F_z))/2,$$
$$D_{xz}(F) = (-E(F_x, F_z) + E(F_x, -F_z) + E(F_z) - E(-F_z))/2.$$

In Eqs. (5.1.2)–(5.1.6) the energies $E(F_\alpha)$ are usually calculated accounting for the BSSE correction using the CP scheme of Boys and Bernardi [45]. Therefore, the

CP-corrected induced first hyperpolarizability of a complex AB in approximation of rigid interacting monomers takes the form:

$$\Delta\beta_{\alpha\beta\gamma} = \beta_{\alpha\beta\gamma}^{AB}(AB) - \beta_{\alpha\beta\gamma}^{A}(AB) - \beta_{\alpha\beta\gamma}^{B}(AB). \qquad (5.1.7)$$

Here $\beta_{\alpha\beta\gamma}^{AB}(AB)$ is the first hyperpolarizability of the complex AB calculated using thee dimer basis set AB. And respectively $\beta_{\alpha\beta\gamma}^{A}(AB)$ and $\beta_{\alpha\beta\gamma}^{B}(AB)$ are the first hyperpolarizabilities of the molecules A and B.

5.1.2 Analytical Representation

In the manner described above for dipole moments and polarizabilities of two interacting molecules A and B the first dipole hyperpolarizability for this system can be written in the form

$$\beta_{\alpha\beta\gamma} \equiv \beta_{\alpha\beta\gamma}^{AB} = \beta_{\alpha\beta\gamma}^{A} + \beta_{\alpha\beta\gamma}^{B} + \beta_{\alpha\beta\gamma}^{ind} + \beta_{\alpha\beta\gamma}^{disp}, \qquad (5.1.8)$$

where $\beta_{\alpha\beta\gamma}^{A}$ and $\beta_{\alpha\beta\gamma}^{B}$ are the dipole hyperpolarizabilities of free molecules A and B; and $\beta_{\alpha\beta\gamma}^{ind}$ and $\beta_{\alpha\beta\gamma}^{disp}$ are the induction and dispersion contributions to the first hyperpolarizability of interacting molecules. In (5.1.8) the contribution to the first hyperpolarizability, caused by interaction of two molecules, is expressed as

$$\Delta\beta_{\alpha\beta\gamma} = \beta_{\alpha\beta\gamma}^{ind} + \beta_{\alpha\beta\gamma}^{disp}. \qquad (5.1.9)$$

Then, the use of (5.1.2), (2.3.4) and (2.3.5) allows us to obtain the analytical expressions for induction and dispersion contributions to the first hyperpolarizability. As a result, the induction part $\beta_{\alpha\beta\gamma}^{A,ind}$ for uncharged molecule A restricted, for instance, by the terms through $\sim R^{-4}$ takes the form

$$\begin{aligned}
\beta_{\alpha\beta\gamma}^{A,ind} = & \alpha_{\alpha\delta}^{B} T_{\delta\varepsilon} \beta_{\varepsilon\beta\gamma}^{B} + \beta_{\alpha\beta\delta}^{A} T_{\delta\varepsilon} \alpha_{\varepsilon\gamma}^{B} + \beta_{\alpha\gamma\delta}^{A} T_{\delta\varepsilon} \alpha_{\varepsilon\beta}^{B} + \gamma_{\alpha\beta\gamma\delta}^{A} T_{\delta\varepsilon} \mu_{\varepsilon}^{B} \\
& + \frac{1}{3}\alpha_{\alpha\rho}^{A} T_{\rho\sigma\varepsilon} B_{\beta\gamma,\sigma\varepsilon}^{B} - \frac{1}{3} B_{\alpha\beta,\delta\varepsilon}^{A} T_{\delta\varepsilon\varphi} \alpha_{\varphi\gamma}^{B} - \frac{1}{3} B_{\alpha\gamma,\delta\varepsilon}^{A} T_{\delta\varepsilon\varphi} \alpha_{\varphi\beta}^{B} \\
& + \frac{1}{3}\beta_{\alpha\beta\rho}^{A} T_{\rho\sigma\varepsilon} A_{\gamma,\sigma\varepsilon}^{B} + \frac{1}{3}\beta_{\alpha\gamma\rho}^{A} T_{\rho\sigma\varepsilon} A_{\beta,\sigma\varepsilon}^{B} + \frac{1}{3}\gamma_{\alpha\beta\gamma\rho}^{A} T_{\rho\sigma\varepsilon}\Theta_{\sigma\varepsilon}^{B} + 6N_{\alpha,\beta,\gamma,\rho\sigma}^{A} T_{\rho\sigma\varepsilon} \mu_{\varepsilon}^{B} + \cdots.
\end{aligned}$$

$$(5.1.10)$$

The full induction part of $\beta_{\alpha\beta\gamma}$ is determined as

$$\beta_{\alpha\beta\gamma}^{ind} = (1 + P^{AB})\beta_{\alpha\beta\gamma}^{A,ind}. \qquad (5.1.11)$$

The dispersion part of hyperpolarizability $\beta_{\alpha\beta\gamma}^{disp}$ of two interacting molecules, following the works [46–48], may be written in the form (restricted by the leading term $\sim R^{-6}$)

$$
\beta_{\alpha\beta\gamma}^{disp} = \frac{1}{12\pi} S(\alpha,\beta,\gamma) \left[T_{\rho\sigma} T_{\varepsilon\eta} \int_0^\infty d\omega\, \alpha_{\sigma\varepsilon}^B(i\omega) \delta_{\eta\rho\alpha\beta\gamma}^A(i\omega,0,0,0) \right.
$$

$$
\left. + 3 T_{\rho\sigma} T_{\varepsilon\eta} \int_0^\infty d\omega\, \gamma_{\sigma\varepsilon\alpha\beta}^B(i\omega,0,0) \beta_{\eta\rho\gamma}^A(i\omega,0) \right]. \tag{5.1.12}
$$

Here, $\alpha(i\omega)$, $\beta(i\omega,0)$, $\gamma(i\omega,0,0)$, and $\delta(i\omega,0,0,0)$ are the imaginary frequency-dependent dipole polarizability, first, second, and third dipole hyperpolarizabilities of the molecules.

It should be noted that the integrals in (5.1.12) may be easily eastimated using the constant ratio approximation CRA2 [46]. This approximation allows estimating the values of $\beta_{\alpha\beta\gamma}^{disp}$ with accuracy $\sim 25\ \%$ and avoid laborious calculations which may be often impossible due to absence of the accurate $\beta(i\omega,0)$, $\gamma(i\omega,0,0)$, and $\delta(i\omega,0,0,0)$ values as functions of $i\omega$ for free molecules. As a result, the integrals may be written in the following forms [49] (see some definitions used here in Sect. 3.1.2):

$$
\int_0^\infty d\omega\, \alpha_{\sigma\varepsilon}^B(i\omega) \delta_{\eta\rho\alpha\beta\gamma}^A(i\omega,0,0,0) \cong \frac{\pi}{30} C_6 \frac{\alpha_{\sigma\varepsilon}^B(0)\delta_{\eta\rho\alpha\beta\gamma}^A(0,0,0,0)}{\alpha^A\alpha^B} \left[\frac{4 + 15\Delta_2 + 20\Delta_2^2 + 10\Delta_2^3}{(1+\Delta_2)^3} \right],
$$

$$
\tag{5.1.13}
$$

and

$$
\int_0^\infty d\omega \gamma_{\sigma\varepsilon\alpha\beta}^B(i\omega,0,0) \beta_{\eta\rho\gamma}^A(i\omega,0) \cong \frac{\pi}{54} C_6 \frac{\gamma_{\sigma\varepsilon\alpha\beta}^B(0,0,0)\beta_{\eta\rho\gamma}^A(0,0)}{\alpha^A\alpha^B} \left[\frac{9 + 32\Delta_1 + 38\Delta_1^2 + 12\Delta_1^3}{(1+\Delta_1)^3} \right].
$$

$$
\tag{5.1.14}
$$

Then, assuming that $\Delta_1 \cong \Delta_2$, the expression for the first hyperpolarizability $\beta_{\alpha\beta\gamma}^{disp}$ can be expressed as:

$$
\beta_{\alpha\beta\gamma}^{disp} = S(\alpha,\beta,\gamma) C_6 \frac{T_{\rho\sigma} T_{\varepsilon\eta}}{\alpha^A\alpha^B} \left[\frac{49}{2880} \alpha_{\sigma\varepsilon}^B \delta_{\eta\rho\alpha\beta\gamma}^A + \frac{91}{1728} \gamma_{\sigma\varepsilon\alpha\beta}^B \beta_{\eta\rho\gamma}^A \right], \tag{5.1.15}
$$

where the following notations are used: $\alpha_{\alpha\beta} \equiv \alpha_{\alpha\beta}(0)$, $\beta_{\alpha\beta\gamma} \equiv \beta_{\alpha\beta\gamma}(0,0)$, $\gamma_{\alpha\beta\gamma\varepsilon} \equiv \gamma_{\alpha\beta\gamma\varepsilon}(0,0,0)$, and $\delta_{\alpha\beta\gamma\varepsilon\varphi} \equiv \delta_{\alpha\beta\gamma\varepsilon\varphi}(0,0,0,0)$. Here $S(\alpha,\beta,\gamma)$ implies summation of all the terms appeared by permuting indices α, β, and γ.

It should be emphasized again, that the multipole polarizabilities and moments in Eqs. (5.1.10) and (5.1.15) are written there in the laboratory system of coordinates and depend on the mutual orientation of the interacting molecules. As a result, the first dipole hyperpolarizability of two interacting molecules is a function (surface) of several variables: Euler angles (rotation of the first and the second molecule), the intermolecular separation R and the internal coordinates when the molecules are considered as nonrigid ones.

5.2   First Hyperpolarizabilitiy of the CH_4–N_2 van der Waals Complex

The references given in the preamble of this chapter give exhaustive enough representation of hyperplarizabilities for small van der Waals complexes including the atom-atomic and atom-diatomic ones. Here, we dwell in more detail on the CH_4–N_2 van der Waals complex which help us to illustrate the calculation technique to be applied to the first hyperpolarizability of a molecule or any molecular complex. The CH_4–N_2 complex is very important and interesting complex that exists in methane-nitrogen atmosphere of Titan.

The induction term $\beta_{\alpha\beta\gamma}^{ind}$ in Eq. (5.1.11) for the interacting molecules CH_4 and N_2, if the terms up to $\sim R^{-6}$ are taken into account, is written as [49]

$$
\begin{aligned}
\beta_{\alpha\beta\gamma}^{ind} = S(\alpha,\beta,\gamma)\Bigg[&\frac{1}{2}\beta_{\alpha\beta\rho}^{A}T_{\rho\sigma}\alpha_{\sigma\gamma}^{B} + \frac{1}{6}\alpha_{\alpha\rho}^{A}T_{\rho\sigma\varepsilon}B_{\beta,\gamma,\sigma\varepsilon}^{B} - \frac{1}{6}\alpha_{\alpha\rho}^{B}T_{\rho\sigma\varepsilon}B_{\beta,\gamma,\sigma\varepsilon}^{A} + \frac{1}{18}\gamma_{\alpha\beta\gamma\rho}^{A}T_{\rho\sigma\varepsilon}\Theta_{\sigma\varepsilon}^{B} \\
&+ \frac{1}{30}\beta_{\alpha\beta\rho}^{A}T_{\rho\sigma\varepsilon\eta}E_{\gamma,\sigma\varepsilon\eta}^{B} - \frac{1}{18}A_{\alpha,\rho\sigma}^{A}T_{\rho\sigma\varepsilon\eta}B_{\beta,\gamma,\varepsilon\eta}^{B} + \frac{1}{90}\gamma_{\alpha\beta\gamma\rho}^{B}T_{\rho\sigma\varepsilon\eta}\Omega_{\sigma\varepsilon\eta}^{A} \\
&+ \frac{1}{30}M_{\alpha,\beta,\rho\sigma}^{A}T_{\rho\sigma\varepsilon\eta}\alpha_{\eta\gamma}^{B} - \frac{1}{54}N_{\alpha,\beta,\gamma,\rho\sigma}^{A}T_{\rho\sigma\varepsilon\eta}\Theta_{\varepsilon\eta}^{B} + \frac{1}{210}\alpha_{\alpha\rho}^{A}T_{\rho\sigma\varepsilon\eta\varphi}G_{\beta,\gamma,\sigma\varepsilon\eta\varphi}^{B} \\
&- \frac{1}{210}\alpha_{\alpha\rho}^{B}T_{\rho\sigma\varepsilon\eta\varphi}G_{\beta,\gamma,\sigma\varepsilon\eta\varphi}^{A} + \frac{1}{90}E_{\alpha,\rho\sigma\varepsilon}^{A}T_{\rho\sigma\varepsilon\eta\varphi}B_{\beta,\gamma,\eta\varphi}^{B} - \frac{1}{90}E_{\alpha,\rho\sigma\varepsilon}^{B}T_{\rho\sigma\varepsilon\eta\varphi}B_{\beta,\gamma,\eta\varphi}^{A} \\
&+ \frac{1}{630}\gamma_{\alpha\beta\gamma\rho}^{A}T_{\rho\sigma\varepsilon\eta\varphi}\Phi_{\sigma\varepsilon\eta\varphi}^{B} - \frac{1}{630}\gamma_{\alpha\beta\gamma\rho}^{B}T_{\rho\sigma\varepsilon\eta\varphi}\Phi_{\sigma\varepsilon\eta\varphi}^{A} + \frac{1}{270}P_{\alpha,\beta,\gamma,\rho\sigma\varepsilon}^{A}T_{\rho\sigma\varepsilon\eta\varphi}\Theta_{\eta\varphi}^{B} \\
&+ \frac{1}{2}\alpha_{\alpha\rho}^{B}T_{\rho\sigma}\beta_{\sigma\beta\varepsilon}^{A}T_{\varepsilon\varphi}\alpha_{\varphi\gamma}^{B} + \frac{1}{2}\alpha_{\alpha\rho}^{A}T_{\rho\sigma}\alpha_{\sigma\eta}^{B}T_{\eta\varphi}\beta_{\varphi\beta\gamma}^{A}\Bigg]
\end{aligned}
\tag{5.2.1}
$$

where again in accordance with Sect. 2.2.1

$$
\underbrace{T_{\alpha\beta\ldots\nu}}_{n} = T_{\alpha\beta\ldots\nu}^{A\to B} = (-1)^{n}T_{\alpha\beta\ldots\nu}^{B\to A}.
\tag{5.2.2}
$$

As mentioned earlier, the superscripts A and B denote the molecules CH_4 and N_2, respectively. The dispersion contribution to the first hyperpolarizability for this complex is calculated using Eq. (5.1.15).

5.2.1   First Hyperpolarizability Surface

The results of ab initio calculations of the induced first hyperpolarizability tensor $\Delta\beta_{\alpha\beta\gamma}$ as functions of R are given in this section in the same way as before. The six configurations of the CH_4–N_2 complex, which are the most discussed ones by researchers (see [50–55]), are considered here. The configurations of the complex and their parameters are also the same as used in Chap. 3 (Figs. 3.6 and 3.7, and Table 3.3). The ab initio and long-range analytical functions of the independent tensor components of $\Delta\beta_{\alpha\beta\gamma}$ are given in Fig. 5.1. The multipole moments and higher polarizabilities of the CH_4 and N_2 molecules calculated at the CCSD(T)/ aug-cc-pVTZ level of theory [49] were used to calculate the values of $\Delta\beta_{\alpha\beta\gamma}$ by means of Eqs. (5.1.15) and (5.2.1). These molecular parameters are given in Table 5.2 to be compared with known ones (Table 5.1).

The analysis of Fig. (5.1) shows that the analytical description of $\Delta\beta_{\alpha\beta\gamma}$ may be effectively used for $R > 11$ a.u. Naturally, for shorter R the analytical values of $\Delta\beta_{\alpha\beta\gamma}$ are noticeably differ from ab initio values due to exchange interactions appeared in this range. It is also noticeable that the exchange interaction is the most important for the $\Delta\beta_{xxx}$ component. Such behavior of $\Delta\beta_{xxx}$ is consistent with results of the asymptotic model of exchange interactions for van der Waals complexes [54, 60]. The strong anisotropy of the exchange effects was also observed for the case of collision-induced dipole moments and dipole polarizabilities in ion-atom pairs [61].

5.2.2   First Hyperpolarizability of the Most Stable Configuration

As it was noted in Chaps. 3 and 4 the dipole moment modulus and the polarizability tensor invariants practically do not change for the family of the most stable configurations [52]. We can expect, that the first-hyperpolarizability tensor invariants which are important for description of interaction-induced hyper-Rayleigh scattering, also change weakly for all most stable configurations.

The hyper-Rayleigh scattering when the incident light has linear polarization may be described by the two tensor invariants of the quadratic hyperpolarizability [3, 62, 63] (in the general case there are six rotation invariants of β [63])

$$A^2 = \sum_i \beta_{iii}^2 + \sum_{i \neq j} \beta_{iij}^2 + 2\sum_{i \neq j} \beta_{iii}\beta_{ijj} + \sum_{i \neq j \neq k} \beta_{ijj}\beta_{ikk} \qquad (5.2.3)$$

and

$$B^2 = \sum_i \beta_{iii}^2 + \frac{11}{3}\sum_{i \neq j} \beta_{iij}^2 - \frac{2}{3}\sum_{i \neq j} \beta_{iii}\beta_{ijj} - \frac{1}{3}\sum_{i \neq j \neq k} \beta_{ijj}\beta_{ikk} + \frac{4}{3}\sum_{i \neq j \neq k} \beta_{ijk}^2. \qquad (5.2.4)$$

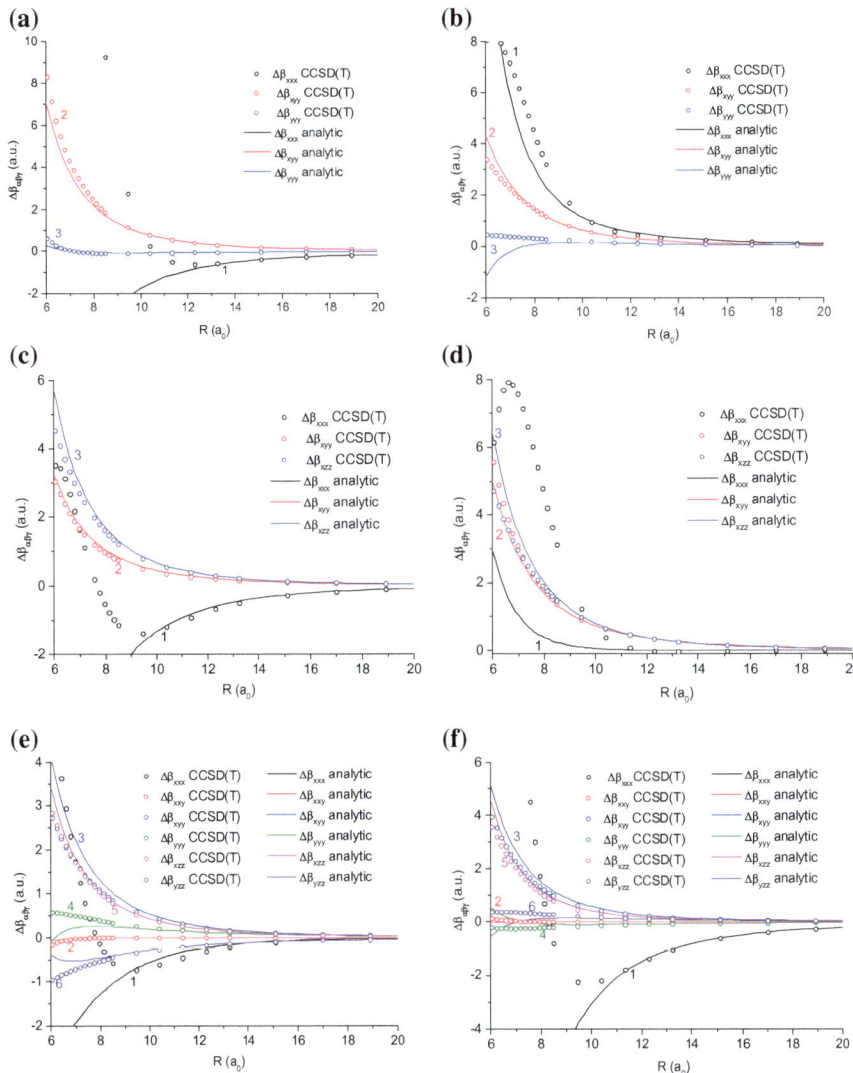

Fig. 5.1 Induced first hyperpolarizability tensor components for six configurations of the CH$_4$–N$_2$ complex

Indeed, the results of ab initio calculations have shown that these invariants for different most stable configurations of the CH$_4$–N$_2$ complex change by less than 0.1 % [49]. The calculated values of the invariants A^2 and B^2 of the complex are presented in Table 5.3. Note, that at $R = \infty$ the invariants A^2 and B^2 are fully determined by $\beta_{\alpha\beta\gamma}$ of the CH$_4$ molecule because for N$_2$ the hyperpolarizability $\beta_{\alpha\beta\gamma}$ is equal to zero.

Table 5.1 The number of nonzero and independent components of $\beta_{\alpha\beta\gamma}$ for general configurations (Fig. 3.6) of the CH$_4$-N$_2$ complex

Configurations: symmetry	Number of nonzero components	Number of independent components	$\beta_{\alpha\beta\gamma}$ [3]
1, 3: C$_{3v}$	11	3	β_{xxx}, $\beta_{yyy} = -\beta_{yzz} = -\beta_{zyz} = -\beta_{zzy}$, $\beta_{xzz} = \beta_{zxz} = \beta_{xxz} = \beta_{xyy} = \beta_{yxy} = \beta_{yyx}$
2, 6: C$_{2v}$	7	3	β_{xxx}, $\beta_{xzz} = \beta_{zxz} = \beta_{xxz}$,
4, 5: C$_s$	14	6	β_{xxx}, β_{yyy}, $\beta_{xyy} = \beta_{yxy} = \beta_{yyx}$, $\beta_{xxy} = \beta_{xyx} = \beta_{yxx}$, $\beta_{xzz} = \beta_{zxz} = \beta_{zzx}$, $\beta_{yzz} = \beta_{zyz} = \beta_{zzy}$

Then, for the hyper-Rayleigh scattering directed at 90° to the propagation of incident light the intensity (I) of the scattered light and its degree of depolarization (ρ) take the form [62]

$$I \sim 7A^2 + 5B^2, \tag{5.2.5}$$

$$\rho = \frac{2B^2}{7A^2 + 3B^2}. \tag{5.2.6}$$

Calculated values of I and ρ are also given in Table 5.3. These values show that under complex formation both the intensity and degree of depolarization are decreased by ~ 11 % and ~ 13 %, respectively.

5.2.3 First Hyperpolarizability of Free Oriented Interacting CH$_4$ and N$_2$ Molecules

To get the hyperpolarizability of free oriented interacting molecules CH$_4$ and N$_2$ (the case of large intermolecular distances) the hyperpolarizability $\beta_{\alpha\beta\gamma}$ should be averaged over Euler angles using Eqs. (5.1.15) and (5.2.1). In this way, taking into account the Kleinman symmetry rules [64] the following nonzero components are kept [49]:

$$\bar{\beta}_{xxx} = \frac{9}{R^4} \left(\bar{B}^A \bar{\alpha}^B - \bar{B}^B \bar{\alpha}^A \right), \tag{5.2.7}$$

$$\bar{\beta}_{xyy} = \bar{\beta}_{xzz} = -\frac{9}{2R^4} \left(\bar{B}^A \bar{\alpha}^B - \bar{B}^B \bar{\alpha}^A \right). \tag{5.2.8}$$

Table 5.2 Multipole moments and high-order polarizabilities of CH_4 and N_2 (in a.u.)

Property	CH_4		N_2	
	[49]	Lit.	[49]	Lit.
Θ_{zz}	0	2.4095[b]	−1.11	−1.1258[a]
Ω_{xyz}	2.57	−7.69[b]	0	−6.75[a]
Φ_{zzzz}	−7.75		−6.84	
α_{xx}	16.51	16.39[b]	10.24	10.2351[a]
α_{zz}	16.51	16.39[b]	14.97	14.8425[a]
β_{xyz}	−9.24	−8.31[b]	0	
γ_{xxxx}	2096	2254[b]	539	807[a]
γ_{xxzz}	771	800[b]	319	322[a]
γ_{zzzz}	2096	2254[b]	1102	1194[a]
δ_{xyzz}	−7707		0	
$A_{x,yz}$	9.25	9.01[c]	0	
$E_{x,xxx}$	−18.84	−18.9[c]	−23.19	−23.42[a]
$E_{z,zzz}$	−18.84	−18.9[c]	27.12	39.59[a]
$B_{xx,xx}$	−282.75	−256[b]	−101.6	−126[a]
$B_{xx,zz}$	141.37	128[b]	54.68	65[a]
$B_{xz,xz}$	−234.39	−219[b]	−121.76	−124[a]
$B_{zz,zz}$	−282.75	−256[b]	−200.28	−216[a]
$M_{xx,xyz}$	−219.18		0	
$M_{xy,xxz}$	−156.29		0	
$D_{x,yzzz}$	−59.42[e]	-52.64[d]	0	0
$G_{xx,xxxx}$	927.78		412.60	
$G_{xx,xxzz}$	−463.89		−461.54	
$G_{xx,zzzz}$	321.88		386.67	
$G_{zz,zzzz}$	927.78		−876.96	
$G_{zz,xxxx}$	321.87		−328.91	
$G_{xy,xyzz}$	−205.75		−269.84	
$G_{xz,xxxz}$	102.87		440.50	
$G_{xz,xzzz}$	102.87		−590.50	
$N_{xxx,yz}$	947.66		0	
$N_{xxy,xz}$	1205.17		0	
$P_{xxy,xxy}$	13816		0	
$P_{xxy,yzz}$	−3272		0	
$P_{xyz,xyz}$	11558		0	
P_{zzzzzz}	14412		0	

[a]Reference [56], [b]Reference [57], [c]Reference [58], [d]Reference [59], [e]Calculated at the CCSD(T)/aug-cc-pVTZ level of theory (not presented in Ref. [49])

Table 5.3 Tensor invariants of the quadratic hyperpolarizability and the parameters of hyper-Rayleigh scattering for CH_4–N_2 complex (at $R_e = 6.8$ a_0) and for pair of free molecules CH_4 and N_2 (at $R = \infty$) [49] (all values are in a. u.)

CH_4–N_2 complex	Free molecules CH_4 and N_2
$A_e^2 = 36.97$	$A_\infty^2 = 0$
$B_e^2 = 557.25$	$B_\infty^2 = 683.02$
$I_e \sim 3045$	$I_\infty \sim 3415$
$\rho_e = 0.5773$	$\rho_\infty = 2/3$

In Eqs. (5.2.7) and (5.2.8) the mean dipole^2-quadrupole polarizabilities $\bar{B}$ and the mean polarizabilities $\bar{\alpha}$ are written as

$$\bar{B}^A = \frac{2}{5}\left(\bar{B}^A_{zzzz} + 2\bar{B}^A_{xzxz}\right),\tag{5.2.9}$$

$$\bar{B}^B = \frac{2}{15}\left(\bar{B}^B_{zzzz} + \bar{B}^B_{xxzz} + 4\bar{B}^B_{xzxz} + 4\bar{B}^B_{xxxx}\right),\tag{5.2.10}$$

$$\bar{\alpha}^B = \frac{1}{3}\left(2\alpha^B_{xx} + \alpha^B_{zz}\right),\quad \text{and } \bar{\alpha}^A = \alpha^A_{zz}.\tag{5.2.11}$$

Equations (5.2.7) and (5.2.8) show, that after averaging, only $B\alpha$ interactions ($\sim R^{-4}$) give contributions to $\bar{\beta}_{xxx}$ and $\bar{\beta}_{xyy} = \bar{\beta}_{xzz}$. As expected, the Eqs. (5.2.7) and (5.2.8) lead immediately to the expressions of the first polarizabilities for two different interacting spherically symmetric atoms A and B [13].

As a result, for free oriented interacting molecules CH_4 and N_2 the invariants A^2 and B^2 (see Eqs. (5.2.3) and (5.2.4)) can be rewritten as a function of R (with accuracy up to the terms $\sim R^{-6}$) [49]

$$\bar{A}^2(R) = \frac{8\left[90(\bar{\alpha}^B)^2 + 7(\Delta\bar{\alpha}^B)^2\right]\left(\beta^A_{xyz}\right)^2}{25R^6}\tag{5.2.12}$$

and

$$\bar{B}^2(R) = 8\left(\beta^A_{xyz}\right)^2 + \frac{8\left[720(\bar{\alpha}^B)^2 + 191(\Delta\bar{\alpha}^B)^2\right]\left(\beta^A_{xyz}\right)^2}{75R^6}$$
$$+ \frac{288\bar{\alpha}^A\bar{\alpha}^B\left(\beta^A_{xyz}\right)^2}{R^6} + \frac{49C_6\delta_{xyzz}\beta^A_{xyz}}{5\bar{\alpha}^A R^6} + \frac{91C_6\bar{\gamma}^B\left(\beta^A_{xyz}\right)^2}{135\bar{\alpha}^A\bar{\alpha}^B R^6}\tag{5.2.13}$$

Here, the anisotropy polarizability $\Delta\bar{a}^B$ and the mean second hyperpolarizability $\bar{\gamma}^B$ are written in the form

$$\Delta\bar{\alpha}^B = \alpha_{zz}^B - \alpha_{xx}^B, \tag{5.2.14}$$

$$\bar{\gamma}^B = \frac{1}{15}\left(3\gamma_{zzzz}^B + 12\gamma_{xxzz}^B + 8\gamma_{xxxx}^B\right). \tag{5.2.15}$$

The invariants $\bar{A}^2(R)$ and $\bar{B}^2(R)$ are used to describe the collision-induced hyper-Rayleigh scattering.

It should be also noted that for the systems having the center of symmetry the dependence like R^{-8} occurs [36–43] for the invariants $\bar{A}^2(R)$ and $\bar{B}^2(R)$. Moreover, for systems of lower symmetry these invariants can go as R^{-3}.

5.3 Multipole Moments and High-Order Polarizabilities of Some Atmospheric and Interstellar Molecules

Despite the fact that recently there are remarkable publications [65–68] that summarize the results of a study of the electrical properties of molecules we should recognize that these studies have been insufficiently represented in the literature.

In this Section the values of multipole moments and high-order polarizabilities of some atmospheric and interstellar molecules [88] are represented.

To calculate the multipole moments and higher polarizabilities of any molecule the finite-field method is effective as it was shown above. However, it should be also noted that there is another approach, based on calculations of the values of the matrix element for the multipole moments, for example, for the dipole moment $\mu_z = \langle\Psi|\hat{\mu}_z|\Psi\rangle/\langle\Psi\mid\Psi\rangle$. Such method has currently implemented in the program CFOUR [89] for HF, MPn methods up to the third order, as well as for Coupled Cluster methods. In Molpro 2012 [90], this approach is implemented for arbitrary order of multipole moments for variational methods as well as for MP2, MP3, QCISD and QCISD(T) methods.

The Tables 5.4, 5.5, 5.6 show the calculation results of some electrical characteristics for H_2, O_2, N_2, CO_2, CO, CN, HCl, HCN, $NaCl$, OH, N_2H^+, CH_4, and H_2O molecules, which are important for astrophysical and atmospheric problems. In the work [88] the calculations were carried out using the finite-field method at the (R) CCSD(T) level of theory with different aVXZ basis sets (X = Q, 5). For these cases, the amplitudes of the applied fields have been chosen as follows: $F_\alpha = 0.0025$ a.u., $F_{\alpha\beta} = 0.0001$ a.u., $F_{\alpha\beta\gamma} = 0.00{,}001$ a.u. and $F_{\alpha\beta\gamma\delta} = 0.000001$ a.u. Multipole moments up to 4th order are presented in Table 5.4. For comparison, in Table 5.4 the other literature data are also given. Table 5.5 presents the calculated and measured values (we have chosen the more reliable ones) of multipole polarizabilities.

Table 5.6 shows the electrical characteristics of low symmetrical molecule H_2O that is the most important atmospheric absorber of the infrared radiation. In the work [88] the calculation for the H_2O molecule was carried out for the r_0 geometry:

Table 5.4 Independent nonzero dipole (μ), quadrupole (Θ), octopole (Ω) and hexadecapole (Φ) electric moments (all values are in a.u.)[a]

Molecule	Bond length	Basis set[a]	μ_z		Θ_{zz}		Ω_{zzz}		Φ_{zzzz}	
H$_2$	$r_{HH} = 1.4487$	aV5Z	–		0.48[b]	**0.46** [69]	–	–	0.32[b]	0.31 [70]
O$_2$	$r_{OO} = 2.29$	aVQZ	–		–0.21[b]	**–0.3(1)** [71]	–	–	4.67[b]	4.48 [72]
N$_2$	$r_{NN} = 2.1066$	aV5Z	–		–1.08[b]	**–1.09** [73]	–	–	–6.81[b]	–6.75 [56]
CO$_2$	$r_{CO} = 2.2117$	aV5Z	–		–3.15[b]	**–3.12** [74]	–	–	–2.08[b]	–1.7 [75]
CO	$r_{CO} = 2.1322$	aV5Z	0.046[b]	**0.048** [76]	–1.46[b]	**–1.44** [77]	3.54[b]	**4.15** [78]	–8.96[b]	**8.15** [78]
CN	$r_{CN} = 2.21$	aVQZ	0.560[b]	**0.57** [79]	0.42[b]		1.21[b]		–2.76[b]	
HCl	$r_{HCl} = 2.41$	aV5Z	0.432[b]	**0.436** [80]	2.69[b]	**2.78** [81]	3.99[b]	4.17 [44]	13.56[b]	14.3 [44]
HCN	$r_{CH} = 2.0135$ $r_{CN} = 2.1792$	aV5Z	–1.188[b]	–1.18 [82]	1.70[b]	1.65 [82]	–9.82[b]	–9.76 [82]	23.27[b]	22.45 [82]
NaCl	$r_{NaCl} = 4.4613$	awCVQZ	–3.537[b]	**–3.53** [83]	6.14[b]	6.26 [84]	–19.40[b]	–20.30 [84]	47.90[b]	50.33 [84]
OH	$r_{OH} = 1.8502$	aV5Z	–0.649[b]	–0.656 [85]	1.29[b]	1.28 [86]	–2.35[b]	–2.39 [86]	4.99[b]	5.10 [86]
N$_2$H$^+$	$r_{NH} = 1.9551$ $r_{NN} = 2.0649$	aVQZ	–1.070[b]		4.41		–14.25[b]		37.66[b]	
CH$_4$	$r_{CH} = 2.0674$	aVQZ	–		–		2.65[c]	2.41[c] [87]	–7.93[b]	–7.69 [87]

[a]For molecules CO, CN, HCl, HCN, NaCl, OH, N$_2$H$^+$ the axis z is directed to the atoms O, C, H, N, Cl, O, and N, respectively. The positive value of CO dipole moment corresponds to its direction from C$^-$ to O$^+$

[b]Calculated in Ref. [88] using the (R)CCSD(T)/aVXZ method (X = Q, 5, see the column 3 for the X value)

[c]Ω_{xyz} for CH$_4$

Table 5.5 The polarizabilities of H_2, O_2, N_2, CO_2, CO, CN, HCl, HCN, NaCl, OH, N_2H^+, CH_4 molecules (in a.u.)

Property	H_2	O_2	N_2	CO_2	CO	CN	HCl	HCN	NaCl	OH	N_2H^+	CH_4
$\alpha_{xx} = \alpha_{yy}$	4.73[a]	8.20[a]	10.33[a]	12.92[a]	11.86[a]	16.24[a]	16.66[a]	13.93[a]	28.44[a]	6.22[a]	7.75[a]	16.61[a]
α_{zz}	6.72[a]	15.09[a]	15.14[a]	27.11[a]	15.45[a]	25.19[a]	18.34[a]	22.25[a]	33.03[a]	8.63[a]	13.91[a]	16.61[a]
α^{b}	5.39[a] *5.42* [70]	10.49[a] *10.59* [91]	11.93[a] *11.74* [92]	17.65[a] *17.69* [75]	13.06[a] *13.09* [93]	19.22[a] *19.28* [94]	17.22[a] *17.39* [95]	16.70[a] *16.74* [96]	29.97[a] *28.70* [84]	7.03[a]	9.80[a]	16.61[a] *17.24* [92]
$A_{x,xz}$	–	–	–	–	−14.80[a] *−15.01* [97]	5.07[a]	3.49[a] *3.75* [44]	−1.49[a]	47.78[a]	−0.20[a]	1.65[a]	–
$A_{z,zz}$	–	–	–	–	−13.70[a] *−13.92* [97]	4.56[a]	14.01[a] *14.0* [44]	−11.71[a]	72.23[a]	−6.28[a]	−5.57[a]	–
$A_{x,yz}$	–	–	–	–	–	–	–	–	–	–	–	9.32[a] *9.01* [58]
$E_{x,xx}$	4.45[a]	20.95[a] *18.81* [72]	−23.01[a] *−23.42* [56]	190.16[a] *187.6* [75]	58.73[a] *60.9* [97]	75.11[a]	22.52[a] *20.0* [44]	100.16[a]	210.43[a]	13.29[a]	54.08[a]	−18.97[a] *−18.9* [58]
$E_{z,zz}$	−1.78[a]	−17.93[a] *−18.06* [72]	38.25[a] *39.59* [56]	−68.62[a] *−68.9* [75]	−37.65[a] *−38.06* [97]	−40.83[a]	2.19[a] *3.0* [44]	−34.23[a]	−95.17[a]	−8.89[a]	−18.60[a]	−18.97[a] *−18.9* [58]
$C_{x,xx}$	4.83[a]	12.84[a] *13.64* [72]	20.20[a] *20.51* [56]	33.93[a] *34.13* [75]	25.41[a] *25.51* [97]	25.71[a]	32.65[a] *35.38* [44]	34.00[a] *34.98* [96]	90.23[a]	5742.7[a]	13.74[a]	37.30[a] *36.77* [87]
$C_{zz,zz}$	6.36[a]	22.30[a] *23.00* [72]	34.82[a] *34.64* [56]	82.26[a] *80.94* [75]	45.94[a] *46.25* [97]	42.72[a]	39.92[a] *41.68* [44]	67.18[a] *68.59* [96]	160.02[a]	832.2[a]	30.20[a]	37.23[a] *36.77* [87]
$C_{xz,xz}$	4.45[a]	19.75[a] *20.48* [72]	27.54[a] *27.20* [56]	53.73[a] *54.81* [75]	37.33[a] *37.61* [97]	32.09[a]	24.38[a] *26.11* [44]	39.89[a] *40.51* [96]	100.70[a]	8.9[a]	17.90[a]	32.69[a] *31.99* [87]

[a]Calculated in Ref. [88]

[b]$\alpha = 1/3(\alpha_{xx} + \alpha_{yy} + \alpha_{zz})$—the mean polarizability of a molecule

Table 5.6 Multipole moments and polarizabilities of the H_2O molecule. The water molecule is in the xz-plane, the atom O lies on the z-axis in its positive direction

Property	Our work [88]	Ref.	Property	Our work [88]	Ref.
μ_z	−0.729	**−0.7296** [98]	$A_{z,yy}$	−3.90	*4.070* [100]
Θ_{xx}	1.96	**1.96** [99]	$A_{x,xz}$	−7.04	*−6.742* [100]
Θ_{yy}	−1.84	**−1.85** [99]	$A_{y,yz}$	−2.41	*−1.786* [100]
Θ_{zz}	−0.13	**−0.10** [99]	$E_{x,xxx}$	−3.51	*−3.273* [100]
Ω_{zxx}	−3.32	*−3.032* [100] *−3.138* [59]	$E_{x,xyy}$	−3.17	*−4.904* [100]
Ω_{zyy}	1.35	*1.259* [59]	$E_{y,yyy}$	0.88	*4.453* [100]
Ω_{zzz}	1.98	*1.879* [59]	$E_{y,yxx}$	-1.34	*−4.013* [100]
Φ_{xxxx}	−1.17	*−1.046* [59]	$E_{z,zzz}$	−2.56	*−2.986* [100]
Φ_{yyyy}	3.98	*3.995* [59]	$E_{z,xxz}$	4.56	*5.646* [100]
Φ_{zzzz}	−4.00	*−3.603* [59]	$C_{xx,xx}$	14.66	*11.22* [59]
Φ_{xxyy}	−3.41	*−3.276* [59]	$C_{yy,yy}$	15.25	*9.44* [59]
Φ_{xxzz}	4.57	*4.322* [59]	$C_{zz,zz}$	14.16	*10.69* [59]
Φ_{yyzz}	−0.57	*−0.719* [59]	$C_{xx,yy}$	−7.88	*−4.98* [59]
α	9.72	**9.64** [101] **9.92** [102]	$C_{xx,zz}$	−6.78	*−6.23* [59]
α_{xx}	10.17	**10.31** [102]	$C_{yy,zz}$	−7.37	*−4.46* [59]
α_{yy}	9.30	**9.55** [102]	$C_{xz,xz}$	13.10	*10.35* [59]
α_{zz}	9.69	**9.91** [102]	$C_{xy,xy}$	10.98	*7.65* [59]
$A_{z,zz}$	−2.65	*−2.194* [100]	$C_{yz,yz}$	10.96	*6.78* [59]

$r_{OH} = 0.9716\ a_0$ and $\angle HOH = 104.68°$. For NaCl molecule the electron correlation of the core and valence-core electrons has also been taken into account using the specially designed awCVQZ basis set. The Tables 5.4, 5.5, 5.6 show a good agreement between experimental (highlighted as a bold type in all tables) and calculated (highlighted as an Italic type in all tables) parameters obtained by different researchers.

References

1. A.D. Buckingham, Permanent and induced molecular moments and long-range intermolecular forces. Adv. Chem. Phys. **12**, 107–142 (1967)
2. A.D. Buckingham, in *Intermolecular Interaction: From Diatomic to Biopolymers*, ed. By B. Pullman (Wiley, New York, 1978), p. 1–68
3. S. Kielich, *Molekularna Optyka Nieliniowa (Nonlinear Molecular Optics)* (Panstwowe Wydawnictwo Naukowe, Warszawa, Poznan, 1977)
4. G. Birnbaum (ed.), *Phenomena Induced by Intermolecular Interactions*, NATO ASI Ser. B 127 (Plenum, New York, 1985)
5. L. Frommhold, *Collision-Induced Absorption in Gases* (Cambridge University Press, Cambridge, 1993)

6. G.C. Tabitz, M.N. Neumann(eds.), *Collision and Interaction-Induced Spectroscopy*, NATO ASI Ser. C 452 (Kluwer, Dordrecht, 1995)
7. A.A. Vigasin, Z. Slanina (eds.), *Molecular Complexes in Earth's Planetary, Cometary, and Interstellar Atmospheres* (World Scientific, Singapore, 1998)
8. C. Camy-Preyt, A. Vigasin (eds.), *Weakly Interacting Molecular Pairs: Unconventional Absorbers of Radiation in the Atmosphere.* NATO ASI, Series IV: Earth and Environmental Sciences, vol 27(Kluwer Academic Publishers, Dordrecht, 2003)
9. J.M. Hartmann, C. Boulet, D. Robert, *Collisional Effects on Molecular Spectra: Laboratory Experiments and Models, Consequences for Applications* (Elsevier, Amsterdam, 2008)
10. K.L.C. Hunt, Long-range dipoles, quadrupoles, and hyperpolarizabilities of interacting inert-gas atoms. Chem. Phys. Lett. **70**(2), 336–342 (1980)
11. K.L.C. Hunt, in *Phenomena Induced by Intermolecular Interactions*, ed. By G. Birnbaum. NATO ASI Ser. B 127 (Plenum, New York. 1985), pp. 263–290
12. A.D. Buckingham, E.P. Concannon, I.D. Hands, Hyperpolarizability of interacting atoms. J. Phys. Chem. **98**(41), 10455–10459 (1994)
13. X. Li, K.L.C. Hunt, J. Pipin, D.M. Bishop, Long-range, collision-induced hyperpolarizabilities of atoms or centrosymmetric linear molecules: theory and numerical results for pairs containing H or He. J. Chem. Phys. **105**(24), 10954–10968 (1996)
14. T. Bancewicz, Interaction-induced pair hyperpolarizabilities by spherical irreducible tensors. J. Chem. Phys. **111**(16), 7440–7445 (1999)
15. T. Bancewicz, Asymptotic multipolar expansion of collision-induced properties. J. Chem. Phys. **134**(10), 104309 (2011)
16. C.E. Dykstra, S.-Y. Liu, D.J. Malik, The hydrogen bonding influence on polarizability and hyperpolarizability. A derivative hartree-fock study of the electrical properties of hydrogen fluoride and the hydrogen fluoride dimer. J. Mol. Struct. THEOCHEM **135**(1), 357–368 (1986)
17. M.G. Papadopoulos, J. Waite, On the interaction hyperpolarisability of He_2, He_3 and Ne_2. An *ab initio* study. Chem. Phys. Lett. **135**(4–5), 361–366 (1987)
18. K.S. Kim, B.J. Mhin, U.-S. Choi, K. Lee, Ab initio studies of the water dimer using large basis sets: the structure and thermodynamic energies. J. Chem. Phys. **97**(9), 6649–6662 (1992)
19. G. Maroulis, Static hyperpolarizability of the water dimer and the interaction hyperpolarizability of two water molecules. J. Chem. Phys. **113**(5), 1813–1820 (2000)
20. G. Maroulis, Computational aspects of interaction hyperpolarizability calculations. A study on $H_2 \cdots H_2$, Ne$\cdots$HF, Ne$\cdots$FH, He$\cdots$He, Ne$\cdots$Ne, Ar$\cdots$Ar, and Kr$\cdots$Kr. J. Phys. Chem. A **104**(20), 4772–4779 (2000)
21. G. Maroulis, A. Haskopoulos, Interaction induced (hyper)polarizability in Ne$\cdots$Ar. Chem. Phys. Lett. **358**(1–2), 64–70 (2002)
22. B.-Q. Wang, Z.-R. Li, D. Wu, C.-C. Sun, *Ab initio* study of the interaction hyperpolarizabilities of the van der Waals complex Ar–HF. J. Mol. Struc. **620**(1), 77–86 (2003). (THEOCHEM)
23. B.-Q. Wang, Z.-R. Li, D. Wu, X.-Y. Hao, R.-J. Li, C.-C. Sun, *Ab initio* study of the interaction hyperpolarizabilities of H-bond dimers between two π-systems. J. Phys. Chem. A **108**(13), 2464–2468 (2004)
24. J. López Cacheiro, B. Fernández, D. Marchesan, S. Coriani, C. Hättig, A. Rizzo, Coupled cluster calculations of the ground state potential and interaction induced electric properties of the mixed dimers of helium, neon and argon. Mol. Phys. **102**(1), 101–110 (2004)
25. D. Wu, Z.-R. Li, Y.-H. Ding, M. Zhang, Z.-R. Zheng, B.-Q. Wang, X.-Y. Hao, *Ab initio* determination of the interaction hyperpolarizability for the H-bond complex NH_3–HF. J. Comput. Meth. Sci. Eng. **4**, 301–306 (2004)
26. A. Haskopoulos, D. Xenides, G. Maroulis, Interaction dipole moment, polarizability and hyperpolarizability in the KrXe heterodiatom. Chem. Phys. **309**(2–3), 271–275 (2005)
27. T. Bancewicz, G. Maroulis, Rotationally adapted studies of ab initio–computed collision-induced hyperpolarizabilities: the H_2–Ar pair. Phys. Rev. A **79**(4), 042704 (2009)

28. A. Haskopoulos, G. Maroulis, Interaction electric hyperpolarizability effects in weakly bound $H_2O\cdots Rg$ (Rg = He, Ne, Ar, Kr and Xe) complexes. J. Phys. Chem. A **114**(33), 8730–8741 (2010)
29. A. Chantzis, G. Maroulis, Interaction-induced electric properties in Kr–Ne from ab initio and DFT calculations. Is there a discrepancy between theory and experiment for the dipole moment? Chem. Phys. Lett. **507**(1–3), 42–47 (2011)
30. G. Maroulis, Interaction-induced electric properties. in *Chemical Modelling; Applications and Theory*. vol. 9, ed. By M. Springborg (The Royal Society of Chemistry, 2012), pp. 25–60
31. H. Reis, M.G. Papadopoulos, I. Boustani, DFT calculations of static dipole polarizabilities and hyperpolarizabilities for the boron clusters B_n (n = 3–8, 10). Int. J. Quant. Chem. **78**(2), 131–135 (2000)
32. B. Skwara, W. Bartkowiak, A. Zawada, R.W. Góra, J. Leszczynski, On the cooperativity of the interaction-induced (hyper)polarizabilities of the selected hydrogen-bonded trimers. Chem. Phys. Lett. **436**(1–3), 116–123 (2007)
33. B. Skwara, A. Zawada, W. Bartkowiak, On the many-body components of interaction-induced electric properties: linear fluoroacetylene trimer as a case study. Compt. Lett. **3**(2–4), 175–182 (2007)
34. Y.-Z. Lan, Y.-L. Feng, Study of absorption spectra and (hyper)polarizabilities of SiC_n and Si_nC (n = 2–6) clusters using density functional response approach. J. Chem. Phys. **131**(5), 054509 (2009)
35. P. Karamanis, R. Marchal, P. Carbonniére, C. Pouchan, Doping-enhanced hyperpolarizabilities of silicon clusters: A global *ab initio* and density functional theory study of Si_{10} (Li, Na, K)$_n$ (n = 1, 2) clusters. J. Chem. Phys. **135**(4), 044511 (2011)
36. W. Głaz, T. Bancewicz, The hyper-Rayleigh light scattering spectrum of gaseous Ne–Ar mixture. J. Chem. Phys. **118**(14), 6264–6269 (2003)
37. W. Głaz, T. Bancewicz, J.L. Godet, Hyper-Rayleigh spectral intensities of gaseous Kr–Xe mixture. J. Chem. Phys. **122**(22), 224323 (2005)
38. W. Głaz, T. Bancewicz, J.-L. Godet, G. Maroulis, A. Haskopoulos, Hyper-Rayleigh light-scattering spectra determined by ab initio collisional hyperpolarizabilities of He-Ne atomic pairs. Phys. Rev. A **73**(4), 042708 (2006)
39. G. Maroulis, A. Haskopoulos, W. Głaz, T. Bancewicz, J.L. Godet, Collision-induced hyperpolarizability and hyper-Rayleigh spectra in the He–Ar heterodiatom. Chem. Phys. Lett. **428**(1–3), 28–33 (2006)
40. T. Bancewicz, W. Głaz, J.-L. Godet, Moments of hyper-Rayleigh spectra of selected rare gas mixtures. J. Chem. Phys. **127**(13), 134308 (2007)
41. T. Bancewicz, W. Głaz, J.-L. Godet, G. Maroulis, Collision-induced hyper-Rayleigh spectrum of H_2–Ar gas mixture. J. Chem. Phys. **129**(12), 124306 (2008)
42. J.-L. Godet, T. Bancewicz, W. Głaz, G. Maroulis, A. Haskopoulos, Binary rototranslational hyper-Rayleigh spectra of H_2–He gas mixture. J. Chem. Phys. **131**(20), 204305 (2009)
43. T. Bancewicz, J.-L. Godet, G. Maroulis, Collision-induced hyper-Rayleigh spectrum of octahedral molecules: the case of SF_6. J. Chem. Phys. **115**(18), 8547–8551 (2001)
44. G. Maroulis, A systematic study of basis set, electron correlation, and geometry effects on the electric multipole moments, polarizability, and hyperpolarizability of HCl. J. Chem. Phys. **108**(13), 5432–5448 (1998)
45. S.F. Boys, F. Bernardi, The calculations of small molecular interaction by the difference of separate total energies—some procedures with reduced error. Mol. Phys. **19**, 553–566 (1970)
46. X. Li, K.L.C. Hunt, J. Pipin, D.M. Bishop, Long-range, collision-induced hyperpolarizabilities of atoms or centrosymmetric linear molecules: Theory and numerical results for pairs containing H or He. J. Chem. Phys. **105**(24), 10954–10968 (1996)
47. H.B. Callen, T.A. Welton, Irreversibility and generalized noise. Phys. Rev. **83**(1), 34–40 (1951)
48. L.D. Landau, E.M. Lifshitz, *Statistical Physics* (Pergamon, Oxford, 1980)

49. Yu.N. Kalugina, M.A. Buldakov, V.N. Cherepanov, Static hyperpolarizability of the van der Waals complex CH_4–N_2. J. Comput. Chem. **33**(32), 2544–2553 (2012)
50. H. Schindler, R. Vogelsang, V. Staemmler, M.A. Siddiqi, P. Svejda, *Ab initio* intermolecular potentials of methane, nitrogen methane + nitrogen and their use in Monte Carlo simulations of fluids and fluid mixtures. Mol. Phys. **80**(6), 1413 (1993)
51. M. Shadman, S. Yeganegi, F. Ziaie, *Ab initio* interaction potential of methane and nitrogen. Chem. Phys. Lett. **467**, 237 (2009)
52. Y.N. Kalugina, V.N. Cherepanov, M.A. Buldakov, N. Zvereva-Loëte, V. Boudon, Theoretical investigation of the potential energy surface of the van der Waals complex CH_4–N_2. J. Chem. Phys. **131**, 134304 (2009)
53. X. Li, M.H. Champagne, K.L.C. Hunt, Long-range, collision-induced dipoles of T_d –$D_{\infty h}$ molecule pairs: theory and numerical results for CH_4 or CF_4 interacting with H_2, N_2, CO_2, or CS_2. J. Chem. Phys. **109**(19), 8416 (1998)
54. N. Zvereva-Loëte, YuN Kalugina, V. Boudon, M.A. Buldakov, V.N. Cherepanov, Dipole moment surface of the van der Waals complex CH_4–N_2. J. Chem. Phys. **133**(18), 184302 (2010)
55. M.A. Buldakov, V.N. Cherepanov, YuN Kalugina, N. Zvereva-Loëte, V. Boudon, Static polarizability surfaces of the van der Waals complex CH_4–N_2. J. Chem. Phys. **132**(16), 164304 (2009)
56. G. Maroulis, Accurate electric multipole moment, static polarizability and hyperpolarizability derivatives for N_2. J. Chem. Phys. **118**(6), 2673–2687 (2003)
57. G. Maroulis, Electric dipole hyperpolarizability and quadrupole polarizability of methane from finite-field coupled cluster and fourth-order many-body perturbation theory calculations. Chem. Phys. Lett. **226**(3–4), 420–426 (1994)
58. G. Maroulis, Dipole-quadrupole and dipole-octupole polarizability for CH_4 and CF_4. J. Chem. Phys. **105**(18), 8467–8468 (1996)
59. C. Huiszoon, *Ab initio* calculations of multipole moments, polarizabilities and isotropic long-range coefficients for dimethylether, methanol, methane, and water. Mol. Phys. **58**, 865 (1986)
60. M.A. Buldakov, V.N. Cherepanov, Asymptotic model of exchange interactions for polarizability calculation of van der Waals complexes. J. Comp. Meth. Sci. Eng. **10**, 1–16 (2010)
61. P.W. Fowler, A.J. Sadlej, Long-range and overlap effects on collision-induced properties. Mol. Phys. **77**(4), 709–725 (1992)
62. S.J. Syvin, J.E. Rauch, J.C. Decius, Theory of hyper-Raman effects (nonlinear inelastic light scattering): selection rules and depolarization rations for the second-order polarizability. J. Chem. Phys. **43**(11), 4083–4095 (1965)
63. V. Ostroverkhov, R.G. Petschek, K.D. Singer, L. Sukhomlinova, R.J. Twieg, S.-X. Wang, L.C. Chien, Measurements of the hyperpolarizability tensor by means of hyper-Rayleigh scattering. J. Opt. Soc. Am. **17**(9), 1531–1542 (2000)
64. P.A. Kleinman, Nonlinear dielectric polarization in optical media. Phys. Rev. **126**(6), 1977–1979 (1962)
65. G. Maroulis (ed.), *Atoms, Molecules and Clusters in Electric Fields. Theoretical Approaches to Calculation of Electric Polarizability*. Computational, numerical and mathematical methods in science and engineering, vol. 1 (Imperial College Press, Singapore, 2006)
66. G. Maroulis, *Computational Aspects of Electric Polarizability Calculations: Atoms, Molecules and Clusters* (IOS Press, Amsterdam, 2006)
67. G. Maroulis, T. Bancewicz, B. Champagne and A.D. Buckingham (eds.), *Atomic and Molecular Nonlinear Optics: Theory, Experiment and Computation. A Homage to the Pioneering Work of Stanislaw Kielich (1925–1993)* (IOS Press Inc., Amsterdam, 2011)
68. U. Hohm, Experimental static dipole-dipole polarizabilities of molecules. J. Mol. Struct. **1054–1055**, 282–292 (2013)
69. A.D. Buckingham, J.E. Cordle, Nuclear motion corrections to some electric and magnetic properties of diatomic molecules. Mol. Phys. **28**(4), 1037–1047 (1974)

70. P.E.S. Wormer, H. Hettema, A.J. Thakkar, Intramolecular bond length dependence of the anisotropic dispersion coefficients for H_2–rare gas interactions. J. Chem. Phys. **98**(9), 7140–7144 (1993)

71. A.D. Buckingham, R.L. Disch, D.A. Dummur, The quadrupole moments of some simple molecules. J. Am. Chem. Soc. **90**(12), 3104–3107 (1968)

72. M. Bartolomei, E. Carmona-Novillo, M.I. Hernández, J. Campos-Martínez, R. Hernández-Lamoneda, Long-range interaction for dimers of atmospheric interest: dispersion, induction and electrostatic contributions for O_2–O_2, N_2–N_2 and O_2–N_2. J. Comput. Chem. **32**(2), 279–290 (2011)

73. A.D. Buckingham, C. Graham, J.H. Williams, Electric field-gradient-induced birefringence in N_2, C_2H_6, C_3H_6, Cl_2, N_2O and CH_3F. Mol. Phys. **49**(3), 703–710 (1983)

74. H. Kling, W. Huettner, The temperature dependence of the Cotton-Mouton effect of N_2, CO, N_2O, CO_2, OCS, and CS_2 in the gaseous state. Chem. Phys. **90**(1–2), 207–214 (1984)

75. G. Maroulis, Electric (hyper)polarizability derivatives for the symmetric stretching of carbon dioxide. Chem. Phys. **291**(1), 81–95 (2003)

76. W.L. Meerts, F.H. De Leeuw, A. Dymanus, Electric and magnetic properties of carbon monoxide by molecular-beam electric-resonance spectroscopy. Chem. Phys. **22**(2), 319–324 (1977)

77. J.M.M. Roco, A. Calvo Hernandez, S. Velasco, Far-infrared permanent and induced dipole absorption of diatomic molecules in rare-gas fluids. I. Spectral theory. J. Chem. Phys. **103**(21), 9161–9174 (1995)

78. J.M.M. Roco, A. Medina, A. Calvo Hernandez, S. Velasco, Far-infrared permanent and induced dipole absorption of diatomic molecules in rare-gas fluids. II. Application to the CO–Ar system. J. Chem. Phys. **103**(21), 9175–9186 (1995)

79. R. Thomson, F.W. Dalby, Experimental determination of the dipole moments of the $X(^2\Sigma^+)$ and $B(^2\Sigma^+)$ states of the CN molecule. Can. J. Phys. **46**(24), 2815–2819 (1968)

80. E.W. Kaiser, Dipole moment and hyperfine parameters of $H^{35}Cl$ and $D^{35}Cl$. J. Chem. Phys. **53**(5), 1686–1703 (1970)

81. F.H. De Leeuw, A. Dymanus, Magnetic properties and molecular quadrupole moment of HF and HCl by molecular-beam electric-resonance spectroscopy. J. Mol. Spect. **48**(3), 427–445 (1973)

82. C. Pouchan, G. Maroulis, Accurate electric multipole moments for HCN and HCP from CCSD(T) calculations with large Gaussian basis sets. Theor. Chim. Acta. **93**(3), 131–140 (1996)

83. A.J. Hebert, F.J. Lovas, C.A. Melendres, C.D. Hollowell, T.L. Story Jr., K. Street Jr., Dipole moments of some alkali halide molecules by the molecular beam electric resonance method. J. Chem. Phys. **48**(6), 2824–2825 (1968)

84. G. Maroulis, Evaluating the performance of DFT methods in electric property calculations: sodium chloride as a test case. Rep. Theoretical Chem. **2**(1), 1–8 (2013)

85. W.L. Meerts, A. Dymanus, Electric dipole moments of OH and OD by molecular beam electric resonance. Chem. Phys. Lett. **23**(1), 45–47 (1973)

86. L. Laaksonen, F. Muller-Plathe, G.H.F. Diercksen, Fully numerical restricted Hartree-Fock calculations on open-shell hydrides: on the basis-set truncation error. J. Chem. Phys. **89**(8), 4903–4908 (1988)

87. G. Maroulis, Electric dipole hyperpolarizability and quadrupole polarizability of methane from finite-field coupled cluster and fourth-order many-body perturbation theory calculations. Chem. Phys. Lett. **226**(3–4), 420–426 (1994)

88. Yu.N. Kalugina, V.N. Cherepanov, Multipole electric moments and higher polarizabilities of molecules: the methodology and some results of ab initio calculations. Atmos. Oceanic Opt **28**(5), 406–414 (2015)

89. CFOUR, a quantum chemical program package written by J.F. Stanton, J. Gauss, M.E. Harding, P.G. Szalay with contributions from A.A. Auer, R.J. Bartlett, U. Benedikt, C. Berger, D.E. Bernholdt, Y.J. Bomble, L. Cheng, O. Christiansen, M. Heckert, O. Heun, C. Huber, T.-C. Jagau, D. Jonsson, J. Jusélius, K. Klein, W.J. Lauderdale, D.A. Matthews, T.

Metzroth, L.A. Mück, D.P. O'Neill, D.R. Price, E. Prochnow, C. Puzzarini, K. Ruud, F. Schiffmann, W. Schwalbach, C. Simmons, S. Stopkowicz, A. Tajti, J. Vázquez, F. Wang, J. D. Watts and the integral packages MOLECULE (J. Almlöf and P.R. Taylor), PROPS (P.R. Taylor), ABACUS (T. Helgaker, H.J. Aa. Jensen, P. Jørgensen, and J. Olsen), and ECP routines by A. V. Mitin and C. van Wüllen. For the current version, see http://www.cfour.de

90. H.-J. Werner, P.J. Knowles, G. Knizia, F.R. Manby, M. Schütz, P. Celani, T. Korona,R. Lindh, A. Mitrushenkov, G. Rauhut , K.R. Shamasundar, T.B. Adler, R.D. Amos, A. Bernhardsson, A. Berning, D.L. Cooper, M.J. Deegan, A. J. ODobbyn, E. Eckert FGoll, C. Hampel, A. Hesselmann, G. Hetzer, T. Hrenar, G. Jansen, C. Köppl, Y. Liu, A.W. Lloyd, R. A. Mata, A.J. May, S.J. McNicholas, W. Meyer, M.E. Mura, A. Nicklass, D.P. O'Neill, P. Palmieri, D. Peng, K. Pflüger, R. Pitzer, M. Reiher, T. Shiozaki, H. Stoll, A.J. Stone, R. Tarroni, T. Thorsteinsson, M. Wang, A. Wolf, in *Molpro, version 2012.1, a package of ab initio programs*. See http://www.molpro.net

91. A.C. Newell, R.C. Baird, Absolute determination of refractive indices of gases at 47.7 Gigahertz. J. Appl. Phys. **36**(12), 3751–3759 (1965)

92. J.W. Schmidt, M.R. Moldover, Dielectric permittivity of eight gases measured with cross capacitors. Int. J. Thermophys. **24**(2), 375–403 (2003)

93. G.A. Parker, R.T. Pack, Van der Waals interactions of carbon monoxide. J. Chem. Phys. **64** (5), 2010–2012 (1976)

94. M. Medved, M. Urban, V. Kello, G.H.F. Diercksen, Accuracy assessment of the ROHF-CCSD(T) calculations of static dipole polarizabilities of diatomic radicals: O_2, CN, and NO. J. Mol. Struct. **547**(1–3), 219–232 (2001). (Theochem)

95. A. Kumar, W.J. Meath, Integrated dipole oscillator strength and dipole properties for Ne, Ar, Kr, Xe, HF, HCl and HBr. Can. J. Chem. **63**(7), 1616–1630 (1985)

96. G. Maroulis, C. Pouchan, Molecules in static electric fields: Linear and nonlinear polarizability of HCN and HCP. Phys. Rev. A **57**(4), 2440–2447 (1998)

97. G. Maroulis, Quadrupole polarizability and hyperpolarizability of carbon monoxide. Theor. Chim. Acta **84**(3), 245–253 (1992)

98. T.R. Dyke, J.S. Muenter, Electric dipole moments of low J states of H_2O and D_2O. J. Chem. Phys. **59**(6), 3125–3127 (1973)

99. J. Verhoeven, A. Dymanus, Magnetic Properties and molecular quadrupole tensor of the water molecule by Beam-Maser Zeeman Spectroscopy. J. Chem. Phys. **52**(6), 3222–3233 (1970)

100. I.G. John, G.B. Bacskay, N.S. Hush, Finite field method calculations. VI. Raman scattering activities, infrared absorption intensities and higher-order moments: SCF and CI calculations for the isotopic derivatives of H_2O and SCF calculations for CH_4. Chem. Phys. **51**(1–2), 49–60 (1980)

101. C.D. Zeiss, W.J. Meath, Dispersion energy constants C_6(A, B), dipole oscillator strength sums and refractivities for Li, N, O, H_2, N_2, O_2, NH_3, H_2O, NO and N_2O. Mol. Phys. **33**(4), 1155–1176 (1977)

102. A. Weber (ed.), *Raman spectroscopy of Gases and Liquids* (Springer, Berlin, 1979)

Chapter 6
Conclusion

The comparison of ab initio calculations with results obtained in the long-range approximation shows that the analytical calculations may be effectively used for description of the interacting energy, dipole moment, polarizability and first hyperpolarizability of the considered interacting molecules for $R > \sim 10$ a.u. So, we can conclude that analytical calculations for large R are a good support to ab initio calculations for full description of interaction energy surface and electrical properties of any pair of interacting molecules. For the range of smaller R, within the potential well, the exchange-interaction contributions are become to be significant due to overlapping of electron shells of interacting molecules. The exchange effects are strong anisotropic and give especially large contributions to the component of any electrical property determined only by the component of dipole moment directed along the intermolecular axis. The same situation occurs for induced dipole moments, dipole polarizabilities, first hyperpolarizability of any complex and, as we assume, has also to be fulfilled for other higher polarizabilities. As a result, the following receipt of calculations of electric properties (and energy surfaces) for molecular complexes is seen. For the values of intermolecular distances R within the potential well the accurate methods of quantum chemistry must be used. And for large R, which are usually more then van der Waals radius of interacting molecules, the analytical methods based on the long-range approximation can be effectively used to change the resource-consuming ab initio calculations. The use the analytical calculations allows to solve the computational problem appeared when a fine grid of calculated points for surfaces of any properties is needed. And the necessity to have a lot of the points (for periphery of a surface the number of these points tends to infinity) disappears.

© The Author(s) 2017
V.N. Cherepanov et al., *Interaction-induced Electric Properties of van der Waals Complexes*,
SpringerBriefs in Electrical and Magnetic Properties of Atoms, Molecules, and Clusters,
DOI 10.1007/978-3-319-49032-8_6

Appendix A
Transformation Rules of the Tensor Components Under Coordinate System Rotation

The relation between the tensor components in the coordinate system of the complex $M_{\alpha\beta\gamma\ldots\delta}$ and the tensor components in the monomer coordinate systems $M_{\alpha'\beta'\gamma'\ldots\delta'}$ are expressed by the use the rotation matrix

$$\lambda_{\alpha\beta} = \begin{pmatrix} \cos\phi\cos\theta\cos\chi - \sin\phi\sin\chi & -\cos\phi\cos\theta\sin\chi - \sin\phi\cos\chi & \sin\theta\cos\phi \\ \sin\phi\cos\theta\cos\chi + \cos\phi\sin\chi & -\sin\phi\cos\theta\sin\chi + \cos\phi\cos\chi & \sin\theta\sin\phi \\ -\sin\theta\cos\phi & \sin\theta\sin\chi & \cos\theta \end{pmatrix}$$

Here, the Euler angles ϕ, θ and χ describe the rotations of any monomer coordinate system relative the complex coordinate system.

As a result,

$$M_{\alpha\beta\gamma\ldots\delta} = \lambda_{\alpha\alpha'}\lambda_{\beta\beta'}\lambda_{\gamma\gamma'}\cdots\lambda_{\delta\delta'}M_{\alpha'\beta'\gamma'\ldots\delta'}$$

© The Author(s) 2017
V.N. Cherepanov et al., *Interaction-induced Electric Properties of van der Waals Complexes*,
SpringerBriefs in Electrical and Magnetic Properties of Atoms, Molecules, and Clusters,
DOI 10.1007/978-3-319-49032-8

Appendix B
Nonzero Independent Electric Multipole Moments and (Hyper)polarizabilities for Molecules of Different Symmetry Group[a]

Group	Multipole moments					(Hyper)polarizabilities						
	μ_α	$\Theta_{\alpha\beta}$	$\Omega_{\alpha\beta\gamma}$	$\Phi_{\alpha\beta\gamma\delta}$	$\Xi_{\alpha\beta\gamma\delta\varepsilon}$	$\alpha_{\alpha\beta}$	$\beta_{\alpha\beta\gamma}$	$\gamma_{\alpha\beta\gamma\delta}$	$A_{\alpha,\beta\gamma}$	$E_{\alpha,\beta\gamma\delta}$	$B_{\alpha,\beta,\gamma\delta}$	$C_{\alpha\beta,\gamma\delta}$
C_1	3	5	7	9	11	6	10	15	15	21	30	15
C_i	0	5	0	9	0	6	0	15	0	21	30	15
C_s	2	3	4	5	6	4	6	9	8	11	16	9
C_2	1	3	3	5	5	4	4	9	7	11	16	9
C_{2h}	0	3	0	5	0	4	0	9	0	11	16	9
C_{2v}	1	2	2	3	3	3	3	6	4	6	9	6
D_2	0	2	1	3	2	3	1	6	3	6	9	6
D_{2h}	0	2	0	3	0	3	0	6	0	6	9	6
C_4	1	1	1	3	3	2	2	5	3	5	8	5
S_4	0	1	2	3	2	2	2	5	4	5	8	5
C_{4h}	0	1	0	3	0	2	0	5	0	5	8	5
C_{4v}	1	1	1	2	2	2	2	4	2	3	5	4
D_{2d}	0	1	1	2	1	2	1	4	2	3	5	4
D_4	0	1	0	2	1	2	0	4	1	3	5	4
D_{4h}	0	1	0	2	0	2	0	4	0	3	5	4
C_3	1	1	3	3	3	2	4	5	5	7	10	5
S_6	0	1	0	3	0	2	0	5	0	7	10	5
C_{3v}	1	1	2	2	2	2	3	4	3	4	6	4
D_3	0	1	1	2	1	2	1	4	2	4	6	4
D_{3d}	0	1	0	2	0	2	0	4	0	4	6	4
C_{3h}	0	1	2	1	2	2	2	3	2	3	6	3
C_6	1	1	1	1	1	2	2	3	3	3	6	3
C_{6h}	0	1	0	1	0	2	0	3	0	3	6	3
D_{3h}	0	1	1	1	1	2	1	3	1	2	4	3
C_{6v}	1	1	1	1	1	2	2	3	2	2	4	3
D_6	0	1	0	1	0	2	0	3	1	2	4	3
D_{6h}	0	1	0	1	0	2	0	3	0	2	4	3
T	0	0	1	1	0	1	1	2	1	2	3	2
T_h	0	0	0	1	0	1	0	2	0	2	3	2
T_d	0	0	1	1	0	1	1	2	1	1	2	2
O	0	0	0	1	0	1	0	2	0	1	2	2
O_h	0	0	0	1	0	1	0	2	0	1	2	2
$C_{\infty v}$	1	1	1	1	1	2	2	3	2	2	4	3
$D_{\infty h}$	0	1	0	1	0	2	0	3	0	2	4	3
K_h	0	0	0	0	0	1	0	1	0	0	1	1

[a]A.D. Buckingham, in *Intermolecular Interaction: From Diatomic to Biopolymers*, edited by B. Pullman (Wiley, New York, 1978), pp. 1–68

© The Author(s) 2017
V.N. Cherepanov et al., *Interaction-induced Electric Properties of van der Waals Complexes*,
SpringerBriefs in Electrical and Magnetic Properties of Atoms, Molecules, and Clusters,
DOI 10.1007/978-3-319-49032-8

Glossary

e	Charge of electron ($1e = 1.60217733 \times 10^{-19}$ C)
a_0	Bohr radius ($1a_0 = 5.29177249 \times 10^{-11}$ m)
E_h	Energy unit *hartree*[a]
μ_α	Dipole moment[a], ea_0
$\Theta_{\alpha\beta}$	Quadrupole moment, ea_0^2
$\Omega_{\alpha\beta\gamma}$	Octopole moment, ea_0^3
$\Phi_{\alpha\beta\gamma\delta}$	Hexadecapole moment, ea_0^4
$\Xi_{\alpha\beta\gamma\delta\varepsilon}$	2^5-pole moment, ea_0^5
$\alpha_{\alpha\beta}$	Dipole polarizability[a], $e^2a_0^2E_h^{-1}$
$\beta_{\alpha\beta\gamma}$	First hyperpolarizability[a], $e^3a_0^3E_h^{-2}$
$\gamma_{\alpha\beta\gamma\delta}$	Second hyperpolarizability[a], $e^4a_0^4E_h^{-3}$
$A_{\alpha,\beta\gamma}$	Dipole-quadrupole polarizability
$E_{\alpha,\beta\gamma\delta}$	Dipole-octopole polarizability, $e^2a_0^4E_h^{-1}$
$B_{\alpha\beta,\gamma\delta}$	Dipole-dipole-quadrupole polarizability, $e^3a_0^4E_h^{-2}$
$C_{\alpha\beta,\gamma\delta}$	Quadrupole-quadrupole polarizability, $e^2a_0^4E_h^{-1}$
$M_{\alpha\beta,\gamma\delta\varepsilon}$	Dipole-dipole-octupole polarizability, $e^3a_0^5E_h^{-2}$
$G_{\alpha\beta,\gamma\delta\varepsilon\phi}$	Dipole-dipole-hexadecapole polarizability, $e^3a_0^6E_h^{-2}$
$D_{\alpha,\beta\gamma\delta\varepsilon}$	Dipole-hexadecapole polarizability, $e^2a_0^5E_h^{-1}$
$N_{\alpha\beta\gamma,\delta\varepsilon}$	Dipole-dipole-dipole-quadrupole polarizability, $e^4a_0^5E_h^{-3}$
$P_{\alpha\beta\gamma,\delta\varepsilon\phi}$	Dipole-dipole-dipole-octupole polarizability, $e^4a_0^6E_h^{-3}$

[a]Unit conversions used in the book for some quantities (see also, for example, David P. Shelton, Julia E. Rice, Measurements and Calculations of the Hyperpolarkabilities of Atoms and Small Molecules in the Gas Phase, Chem. Rev., 1994, **94**, No. 1, 3–29):

for energy—1 E_h (or 1 a.u.) = $4.35974434 \times 10^{-18}$ J = 627.509181 kcal mol^{-1} = 219 474.6307 cm^{-1};

for dipole moment—1. a.u. = 2.541 75 Debye (D) (1 D = 10^{-18} esu cm) = 8.478 358 $\times$ 10^{-30} C m ;

for polarizability—1 a.u. = 0.148 19 $\times 10^{-24}$ cm^3 (or 0.148 19 Å^3) = 1.648 778 $\times$ 10^{-41} C^2 m^2 J^{-1};

for first hyperpolarizability—1 a. u. = 3.206 361 $\times 10^{-53}$ C^3 m^3 J^{-2};

for second hyperpolarizability—1 a. u. = 0.623 537 7 $\times 10^{-64}$ C^4 m^4 J^{-3}.

© The Author(s) 2017
V.N. Cherepanov et al., *Interaction-induced Electric Properties of van der Waals Complexes*,
SpringerBriefs in Electrical and Magnetic Properties of Atoms, Molecules, and Clusters,
DOI 10.1007/978-3-319-49032-8